P9-DNY-424

WHY WE DREAM

the transformative power of our nightly journey

ALICE ROBB

An Eamon Dolan Book
Houghton Mifflin Harcourt
BOSTON NEW YORK 2018

hmhco.com

Library of Congress Cataloging-in-Publication Data is available.

ISBN 978-0-544-93121-3

Book design by Lisa Diercks / Endpaper Studio

Printed in the United States of America
DOC 10 9 8 7 6 5 4 3 2 1

I understand why most people regard their dreams as of little importance. They are too light for them, and most people identify the serious with what has weight. Tears are serious; one can collect them in a jar. But a dream, like a smile, is pure air. Dreams, like smiles, fade rapidly.

But what if the face faded away, and the smile remained?

— Susan Sontag, *The Benefactor*

CONTENTS

INTRODUCTION

I SPENT THE SUMMER OF 2011 DIGGING HOLES AND talking about my dreams. Within two weeks, I had blown through the novels I'd taken to the remote Andean village of Nepeña, where I was excavating Moche remains with my classmates and a Peruvian professor. I'd saved most of my suitcase for bulky rain gear and emergency jars of peanut butter; I hadn't anticipated how much time I'd have when my internet access was subject to the whims of an erratic café owner. So when my friend James passed me a beat-up paperback whose cover showed a man's brain being penetrated by a ray of sunlight and a puff of clouds, I willed myself to set my skepticism aside.

As I scanned the table of contents, though, I couldn't help but roll my eyes at chapter titles like "Life Is a Dream" and "Rehearsal for Living." I cringed at the list of exercises: the eerie-sounding "twin bodies technique," the ludicrous "dream lotus and flame technique," the ominous "no body technique." Stephen LaBerge's *Exploring the World of Lucid Dreaming* had all the trappings of a New Age self-help screed, but with the closest English-language bookstore a six-hour bus ride away, I started to read.

"Proverbially, and undeniably, life is short," LaBerge wrote. "To make matters worse, we must spend between a quarter and a half of our lives asleep. Most of us are in the habit of virtually sleepwalking through our dreams. We sleep, mindlessly, through many thousands of opportunities to be fully aware and alive." In what LaBerge called "lucid dreams," a sleeping person could become aware that she was dreaming and—with a little practice—control the plot of the dream. I was hooked.

Most people experience a lucid dream at some point in their lives, but only about 10 to 20 percent have them regularly. For some in that minority, lucid dreaming is so pleasurable that it becomes a hobby or a kind of self-help. Lucid dreams can seem more vivid than reality; they can provide a high as intense as psychedelics and even deliver sexual gratification. (One psychologist claimed to reach orgasm in one-third of her lucid dreams, and measures of vaginal pulse amplitude have shown that women's dream orgasms correspond to real physiological changes.) Others use lucid dreaming to take control of nightmares or rehearse difficult real-life situations. Of all my memories of that summer in Peru—drinking pisco in the desert, finding a mummified baby, unwrapping it under less-than-scientifically-optimal conditions—the one that stands out most is the memory of my first lucid dream.

At nine o'clock, I climbed into the bottom bunk and curled up in my sleeping bag, worn out from physical exertion and the monotony of digging. I set my alarm for five a.m. and drifted off almost immediately, my body too tired to let my mind wander down its usual anxiety-laden paths. And then, the scene changed. It was a summer afternoon—not the Andean summer, with its thin warmth and cloudy nights, but a real summer, the kind of heat so extravagant you jump in the water and dry off in the sun. I soaked up the warmth I'd been craving, treading water in some bucolic pool I'd never seen before. I don't particularly like swimming in real life; I don't like exercising in any form without the distraction of podcasts or Pandora. But this was different—effortless and sensual. I had a heightened awareness of every part of my body, the physicality of the cool water and the bright air and a surreal forest enclosing the pool in magnificent foliage. I woke up euphoric.

The memory had none of the haziness that usually clouds dreams, and the details remain perfectly crisp years later. But I wasn't just elated; the whole thing was also vaguely disturbing. I hadn't been in my sleeping bag in a dusty dormitory in Peru — I had been transported to some faraway place, and I preferred it there. My jaunt in the pool had shaken my sense of what was real, and I couldn't explain it without sounding crazy. All I knew was that I wanted to do it again.

James and I spent the rest of the summer practicing LaBerge's tips. We recounted our previous night's dreams while we scratched the grime off ancient pots. We repeated LaBerge's mantra ad nauseam: "Tonight, I will have a lucid dream." We made up mantras of our own: "Tonight, I will fly to the moon." We learned to recognize the signs that we were dreaming, like finding ourselves flying or meeting dead people. Every couple of hours, we would do what LaBerge called a reality test, asking ourselves if we were awake or asleep — a trick that, once ingrained, LaBerge promised would trigger lucidity.

The bar for what constitutes good conversation may be lower when you spend most of your time scraping the sand with a trowel, but even after I left Peru, even when I had more than four people to talk to, high-speed WiFi, and whole libraries full of books, I couldn't stop thinking about dreams. They were so much fuller, so much more mysterious than I had ever imagined.

I began keeping a dream diary, carefully logging whatever I could remember of my dreams in a spiral-bound leather notebook each morning; I had read that it was important to record something every day, no matter how fragmented or boring. The results were almost immediate. Within weeks, the entries in my journal went from a dutiful *No recollection* or brief, tentative snippets (*I am watching the* Nutcracker? *There is a spider?*) to two or three long, convoluted narratives almost every night. My new night life was every bit as active — and at least as entertaining — as my waking hours, and I was stunned:

I understood that I had been having dreams like this all my life, but I had been promptly forgetting them, letting them fade away as though they had never happened. What adventures had I gone on and then forgotten? What opportunities — to gain new insight or just to take a break from reality — had I missed?

Most new skills — especially those that promise to change how you experience the world — are difficult to learn. Mastering a new language takes years of concentrated study. Meditating requires patience and frequent, sometimes frustrating practice. Gains are incremental, often imperceptible. But improving your dream life can be as simple as increasing the time you devote to thinking about dreams from none at all to a minute or two each day, sparing a pre-bed thought for your intention to remember your dreams or taking a moment to write them down or speak them into your smartphone in the morning. The process is painless; the progress is swift. And the payoff is life-changing. Becoming aware of your dreams is like dipping into a well of otherwise inaccessible fantasies and fears, signs from our subconscious and creative solutions to projects and problems.

IN RECENT YEARS, scientists have discovered how we can improve our dream recall and harness the power of dreams in a systematic fashion. But humans have been wondering about their dreams for millennia. Some scholars believe that our ancestors' earliest artwork — cave paintings — were inspired by their makers' nighttime visions. Dream diaries are among the oldest examples of literature; they have been found in the remains of ancient Greece and medieval Japan.

We live in a world built on dreams. Throughout history and around the globe, dreams have been a source of endless fascination and guidance. We have looked at dreams as prophecies of the future and vestiges of the past, as messages from the divine and from within our own psyches. Dreams allow us to experience things we've lost and things we've never had. In dreams, the paralyzed can move; the blind can see. Doctors have used dreams as a tool in diagnosis; artists have relied on them for inspiration. The dying take comfort in vivid dreams of the past, dreams that blur the boundaries of consciousness and call reality into question. Politicians and mythical heroes have looked to dreams to make decisions and invoked them to justify war. Leaders have used them for good (when Gandhi argued against the constriction of Indians' civil liberties in 1919, he said he had dreamed that the country would observe a strike) and for evil (videotapes released after the September 11 attacks show Osama bin Laden and his followers swapping dreams of pilots, planes, and crashing buildings). Even for the less than 3 percent of the population who claim never to remember their dreams, it is still important to understand them as a potent, overlooked force behind famous works of art, religious conversions, and political events.

Our contemporary neglect of our dream lives is not only a historical anomaly but a particular paradox in our current culture. People are obsessed with hearing the latest research on sleep, even if scientists haven't yet reached a consensus on why we pass out every night. We want to know how screens and modern scheduling affect our sleep patterns. We click on studies warning us that anything less than eight hours of sleep destroys our health, looks, and happiness — or promising that six hours is enough or that some people are fine with just three or four.

Meanwhile, we chart, track, and optimize our time, buying Fitbits and phone apps to count the minutes spent on exercise, work, and hobbies; we suffer from "fear of missing out." Yet in ignoring our dreams, we squander an opportunity to experience adventure and boost our mental health, about five or six years' worth of opportunity (20 to 25 percent of total time asleep) over the course of an average lifetime. Sleep is usually discussed as a means to an end — a tool to ensure the daytime is productive, to improve memory, regulate metabolism, and keep the immune system in order. But as LaBerge asked: "If you must sleep through a third of your life, as it seems you must, are you willing to sleep through your dreams too?"

Until recently, there was no such thing as a science of dreams. For reasons both practical and philosophical, the mysteries of dreaming were relegated to the realms of magic and religion. Dreams don't easily lend themselves to the lab; they are difficult to report in full, and, although a new Japanese scanning device may be able to "read" certain dream motifs, they remain impossible to verify. And the scientists who have chosen to follow their interest in dreams have not always been the kind of strait-laced ambassadors who could have best served the cause. The subject has attracted more than its share of brilliant oddballs — awkward obsessives willing to stake their careers on a puzzle they were unlikely to ever crack. But if the heroes of this story have sometimes strayed outside the bounds of scientific orthodoxy — designing doomed experiments on the telepathic nature of dreams, insisting that dreams could predict the future, conflating their own intuition with evidence — their open-mindedness has also helped them recognize surprising truths. I've come to appreciate how blurry that line can be — how legitimate scientists can entertain improbable ideas and how good ideas can come from unlikely places. Against the advice of some of her colleagues, Harvard psychologist Deirdre Barrett accepted a paper on extrasensory perception for the academic journal she edits, *Dreaming*. "My stance is that what defines scholarly research is the approach and the design," she told me. "It's anti-science to insist on a conclusion."

Thanks to a few lucky breakthroughs in the lab and a recent explosion of sleep research, dreams are finally getting their due, gaining more and more credibility within the sciences. The number of sleep labs in the United States is at an all-time high; it has risen from four hundred in 1998 to more than twenty-five hundred today. We have come to appreciate the importance of sleep for health; people around the world spend more than fifty billion dollars a year on sleep aids, and experts expect the insomnia industry to keep growing. Several universities in the United States have begun offering courses and even entire programs in dreams and dream psychology. Philosophers have homed in on dreams as a nexus for theorizing on the mind-body connection and the nature of consciousness.

New developments in technology have also helped revolutionize the study of dreams, enabling scientists to collect dream reports faster and from more diverse populations than ever before. In the twentieth century, most dream research was carried out on white college students. But over the past few years, people of all ages from around the world have been uploading their dreams to websites like Dreamboard and DreamsCloud, and scientists are beginning to unpack the treasure-troves of data within.

The reasons why we dream have turned out to be just as strange and powerful as anyone might have guessed. Dreams play a crucial role in some of our most important emotional and cognitive systems, helping us form memories, solve problems, and maintain our psychological health.

When we dream, we integrate new pieces of information into our preexisting web of knowledge; the brain sifts through the jumble of recent experience, marking off the most important memories for long-term storage. Dreaming about a new skill helps us master it; practicing a task or a new language in our sleep may be as effective as grinding away in real life.

Dreams have inspired stories that have entertained generations of readers and brought about scientific discoveries that have changed the world. We have dreams to thank for the sewing machine and the periodic table. Too many artists and writers to name — including the likes of Beethoven, Salvador Dalí, Charlotte Brontë, Mary Shelley, and William Styron — credit dreams with some of their most famous creations.

We dream in order to work through our anxieties and prepare for our days; we rehearse for trials and tests, making their real-world counterparts feel more familiar. We confront worst-case scenarios in a no-stakes environment so the actual event feels like a comparative breeze. People who dream about new mazes navigate real ones more efficiently. Students who have nightmares about their exams outperform classmates who don't. Dreaming about traumatic events can help us heal from them. Conversely, mood disorders like depression often involve a disruption of normal dreaming; a mind deprived of REM sleep — when most dreams occur — is prone to breaking down. Suicidal thoughts have been linked to a loss of dreaming or a drop in dream recall.

Dreams can help us become more self-aware; they draw deep-seated anxieties and desires to the surface, forcing us to face up to hopes and fears we haven't acknowledged. They offer a window into our psyches; a dream can be the key to recognizing an emotional problem.

If we fail to take the simple steps to remember and understand our dreams, it is as though we are throwing away a gift from our brains without bothering to open it. Some of the cognitive functions of dreaming — like aiding in memory formation — will go on no matter what, provided we get a normal night's sleep; whether or not we take notice, dreams will help us learn new information and assimilate new experiences into long-term memory.

But if we ignore our dreams, we rob ourselves of some of their most powerful benefits. By paying attention to our dreams, we can access ideas that would otherwise vanish into the night. By tracking them over time, we can gain confidence in nerve-racking situations.

If we go a step further and discuss our dreams with therapists or doctors, we stand to reap another reward: dreams can clue us in to mental and physical issues that might otherwise fly under the radar. And if we go all the way and share our dreams more broadly — with like-minded friends or groups of dream enthusiasts — we can glean an even clearer understanding of their sometimes messy metaphors and symbols. We can become fluent in the language of dreams.

The tradition of lucid dreaming in the West is long, but modern scientists have only just begun to respect and explore it. Though accounts of lucid dreams can be found in the writings of Aristotle and Augustine, it wasn't until the 1970s that scientists figured out how to study the phenomenon, and recently those techniques have borne fruit, showing us the therapeutic power of lucid dreams and the steps that most reliably induce them.

In the course of researching this book, I have experimented with cutting-edge technologies — like a virtual-reality treatment for nightmares — as well as primitive practices that take nothing more than my mind and maybe a pen and paper. I have learned concrete steps I can take to improve dream recall, conquer nightmares, and exert control over the content of dreams. I'll explain which methods have been fully tested and which have worked for me, how I went from remembering dreams only occasionally to remembering them whenever I wanted, and how the dreams I recalled became longer, more vivid, and more lucid.

This is a book about science and history; it's the story of how previous cultures forgot about dreams, and how we are finally rediscovering them. As you learn how rich your inner life is as you sleep, I imagine — I hope — that you will want to remember your own dreams more often and even experiment with lucidity. If I succeed in convincing you that dreams matter, you may find yourself remembering more of them without any special effort; just being curious about your dream life is often enough to improve it. Another easy way to improve your dream recall is to spend a little bit more of your waking time thinking about dreams; reading this book counts toward that end. (People have often told me that after chatting with me about my book, they have unusually vivid dreams.) Good dream recall is a prerequisite for lucid dreaming; if you start keeping a dream journal now, you will have a head start when I explain how to induce lucid dreams later on.

It is an exciting moment to embark on this journey. The questions are age-old, and as researchers have made inroads into this mysterious territory, they have sometimes found themselves treading the same paths as their ancestors. But new research in science and psychology — in sometimes fraught conjunction with ancient and mystical beliefs — is shedding a long-awaited light on the meaning and purpose of dreams.

chapter 1

HOW WE FORGOT ABOUT DREAMS

UNTIL THE NINETEENTH CENTURY, DREAMS WERE thought of in the context of spirituality rather than science. In diverse religious traditions, dreams have been treated as a channel through which ordinary people could experience another world and prophets could divine the will of the gods. The biblical Joseph won his post in the royal court by interpreting Pharaoh's dreams, explaining that the seven fat cows and seven skinny cows represented the coming seven years of plenty and seven years of famine. The azan — the Muslim call to prayer — is said to have been inspired by a dream of one of Muhammad's companions. Muhammad's own dreams gave him solace in moments of doubt and confirmation that he was on the right path. Hindu scripture teaches that dreams contain reliable, if counterintuitive, predictions; losing one's teeth in a dream foreshadows death, while a nightmare of being beheaded is a sign of long life. The birth of the Buddha was supposedly heralded by a dream in which his mother, Queen Maya, saw a white elephant bearing a lotus flower walk around her in a circle and then crawl into her womb.

• • •

DREAMS HAVE OFTEN been valued as a window into the future. In the ancient world, doctors treated them like a kind of magical x-ray, consulting dreams for clues to their patients' prognoses. "Beginnings of diseases and other distempers which are about to visit the body," Aristotle wrote in the fourth century BC, "must be more evident in the sleeping than in the waking state." The Greek physician Hippocrates reveled in the diagnostic power of dreams, taking a literal approach, alleging, for example, that dreams of fast-flowing rivers signified an excess of blood in the body. Several centuries later, Galen claimed to "have saved many people by applying a cure prescribed in a dream." He made a point of interrogating his patients about their dreams, just as he asked about physical symptoms, and he took his own dreams seriously too; he credited his path in life to a dream in which Asclepius—the beloved god of healing and dreams—commanded him to become a surgeon.

The Greek dream god inspired cult-like levels of devotion. For thousands of years—long after the civilization that invented him had collapsed—pilgrims and invalids traveled from all over the Mediterranean to worship at his temple in the city of Epidaurus, to sleep in an inner sanctuary called the *abaton*, and to pray for a diagnostic or healing dream. Relics found at Asclepian sanctuaries—terracotta limbs and heads, a finger with a cancerous lump—testify to the vast powers he was thought to possess. One inscription tells of a man named Lucius who traveled to the Asclepian temple at Rome because of a pain in his chest. There, a dream ordered him to gather ashes from the altar, mix them with wine, and apply the elixir to his side. Another describes a blind soldier who received dream-instructions to make a balm out of honey and the blood of a white rooster and smear it on his eyes.

DREAMS CAN BE so lifelike, their sources so enigmatic, and their aftereffects so potent that supernatural explanations can seem almost logical. Dreams of communing with God or visiting the dead can instill a sense of awe in the most committed atheist and compel the more spiritually inclined to wonder whether they've slipped across some celestial threshold. Dreams can even change our beliefs. A Methodist missionary once complained that his targets "more often became 'serious' about their religion and prayers not as the result of preaching, but most commonly a 'warning in a dream.'"

Some scholars even argue that religion itself has its origins in dreams and our attempts to understand them. Psychologist Kelly Bulkeley and neuroscientist Patrick McNamara believe that people invented religious frameworks as a way to make sense of the innately mystical experience of dreaming. Even ordinary dreams plunge us into alternative worlds, universes with different rules or none at all, where people can morph into monsters and superhuman beings take an intimate interest in personal affairs—worlds much like the ones set out in myths. Visions, whether in sleep or in waking life, propel us on a search for answers. Some research suggests that schizophrenics—whose illness is characterized by hallucinations—are more inclined toward religiosity than the general population.

Dreams are a powerful mechanism for generating god concepts or supernatural agents—intelligent, nonhuman beings who appear to have their own independent will. When psychologists Richard Schweickert and Zhuangzhuang Xi analyzed a sample of dream reports that had been uploaded to the dream-sharing website DreamBank, they found an average of about nine instances per dream of "theory of mind events," in which the dreamer assigned independent agency or inner feelings to a dream-character. ("A vampire was afraid of the head vampire"; an animated corpse "wanted to leave"; "someone was astonished" when the dreamer drove her wheelchair over a desk.) In dreams, people attribute motives and emotions to figures that they have invented, similar to how they guess at the will of spirits or gods.

There are parallels, Bulkeley and McNamara point out, in how people grapple with the meaning of dreams and how they analyze religious texts. "Every time we decide to 'read' a dream, we simultaneously anticipate brooding about the dream's events and images several times throughout the day," McNamara wrote in the magazine *Aeon*. "After all, it is essentially impossible to understand a dream the first time through . . . This same paradoxical interpretive stance also occurs when we read sacred scriptures or listen to religious stories or attempt to interpret our own religious experiences (if we are 'believers')." Waking from a vivid dream, like closing a holy book, is only the beginning of the interpretive process; in neither case do we simply accept the experience at face value. Instead, we revel in the raw power of the memory while knowing that we will come back to it. Soon, we will rehash the text or the dream and parse its meaning, initiating a cycle of "endless exegesis, interpretation and re-interpretation," leading "to new meanings" and even "new ritual procedures."

Neurochemical changes that occur during REM sleep prime our brains to not only generate but also trust extraordinary visions. Dopamine — the neurotransmitter associated with pleasure and reward — surges, as does acetylcholine, a chemical involved in memory formation. Activity spikes in the emotion centers of the brain — the amygdala and the limbic system as a whole. At the same time, the dorsolateral prefrontal cortex — the main area involved in rational thinking and decision-making — powers down, and levels of serotonin and norepinephrine, which are associated with self-control, drop. The result is a perfect chemical canvas for dramatic, psychologically intense visions: the parts of the brain that produce emotion are fired up, while the areas that keep them in check are quiet. "People have always wondered why dreams seem to generate religious ideas so easily," McNamara said. "Dreams have a natural cognitive mechanism to produce this super-natural agent concept."

EVEN IN PERIODS of relative skepticism, dreams were widely thought to have a supernatural origin. At the height of the Enlightenment, rational Westerners still consulted their dreams for guidance and glimpses of the future. "Dream interpreters in early America were as widespread as forgers, purveyors of instant cures, and other hucksters," historian Andrew Burstein wrote in *Lincoln Dreamt He Died: The Midnight Visions of Remarkable Americans from Colonial Times to Freud.* Nonsensical guides to dream interpretation ("It is good to dream of white, purple, pink, or green; brown or black is rather ominous") were printed by the heap. Newspapers published cautionary tales of fools who failed to heed their dreams. New Hampshire's *Freeman's Oracle* told of a young sailor's wife who dreamed of seeing her husband's corpse bobbing in the sea and begged him not to join his captain for dinner on deck; he ignored her warning and drowned.

Dreams weren't just the province of the uneducated or the superstitious. Ezra Stiles, president of Yale College in the eighteenth century, diligently recorded stories of acquaintances' prophetic dreams in his journals. President John Adams and physician Benjamin Rush wrote to each other about their dreams, competing to see who enjoyed a richer dream life. Adams was especially moved by a dream in which he explained to a zoo's worth of animals — lions, elephants, wolves — that they ought to form a "sovereign annimatical government."

The rapid pace of technological change in the nineteenth century only boosted Western interest in the supernatural. Ordinary people, awed by their new ability to travel and communicate across once-unthinkable distances, wondered whether mediums and ghosts were any more fantastic than the railway and the telegraph. In the 1880s, a group of eminent British scholars and philosophers banded together to form the Society for Psychical Research. They collected tales of thought transference and ghost sightings and compiled them into a thousand-page case for the existence of the paranormal. They mailed questionnaires to more than five thousand people asking them to report any dreams in which they had foreseen a death that later came to pass, and concluded that these dreams were too common to pin on coincidence.

Newspapers printed readers' dreams of politics as though expecting them to furnish clues to the country's future. Joseph Pulitzer's *New*

York World announced a nationwide "best dream" competition, challenging the paper's hundreds of thousands of readers to write in with their most spectacular visions. The winning entrant — the "Champion Dreamer," as judged by Nathaniel Hawthorne's son Julian — was a former professor from Maryland who signed his letter J.E.J. Buckey. One night, Buckey wrote, he dreamed that he had shot a stranger and stood by as blood gushed from his victim's neck. The next day, as he was walking to work, still reeling from the dream, he claimed to catch sight of the man from his nightmare. And the man supposedly recognized him too; he turned to Buckey and begged him not to shoot. Buckey was sure he understood what had happened: "We had both dreamed the same dream."

IN THE 1850S, the French physician Louis Alfred Maury became one of the first scientists to attempt to study dreams empirically. Using himself as a guinea pig, he played around with his external environment to see if he could influence his own dreams. He had an assistant tickle his nose with a feather as he slept and dreamed that a mask was being torn from his face. He had someone drip water onto his forehead and dreamed that he was sweating and drinking wine. He came to a radical conclusion: dreams didn't come from the gods but from the world around us.

It would be another century before scientists would appreciate the role of dreams in problem-solving, but in 1892, developmental biologist Charles Child asked two hundred college students if they had ever realized something in a dream that helped them tackle a real-life challenge. About 40 percent of his students said that they had; several claimed to have cracked algebra problems overnight. One recalled an instance back in prep school when a dream had helped him with his homework, delivering a perfectly translated passage of Virgil in the morning.

At the turn of the century, Sigmund Freud elevated dreams to a new status, lending them a degree of academic legitimacy for the first time.

He made dreams the focal point of the new discipline of psychoanalysis, lauding them as the "royal road to a knowledge of the unconscious activities of the mind." "Psychoanalysis is founded upon the analysis of dreams," he declared.

By examining them, he argued in *The Interpretation of Dreams,* patients — or their analysts — could discover their secret wishes and unravel the unconscious, thus empowering them to treat their neuroses. Since dreams arise from our own minds, every part of the dream — strangers, lovers, inanimate objects — symbolizes some aspect of the self.

One of Freud's most radical claims was that dreams represent wish fulfillment; they allow us to satisfy desires that we are aware of as well as wishes we can't acknowledge even to ourselves. The wish may be as profound as a longing to return to childhood and secure the love of an emotionally distant parent or as simple as wanting to alleviate a hunger pang that has arisen in the night. When she was nineteen months old, Freud's daughter Anna threw up after gorging on strawberries and wasn't allowed to eat for the rest of the day. That night, Sigmund heard her cry out in her sleep: "Anna Feud, strawberry, huckleberry, omelette, pap!" In a dream, her father presumed, the baby was satisfying her hunger.

Usually, the wish is not so transparent. Freud believed that the oblique nature of dreams constitutes a layer of protection allowing us to sleep through the night without being overwhelmed by the issues at the dreams' core; like sunglasses shielding retinas from direct sunlight, they insulate us from what we can't handle. Freud distinguished between the dream's "manifest content" (the plots and images as they are remembered) and its "latent content" (the repressed desires that inspire them). Throughout the day, he believed, a mechanism called the "censor" policed the mind, keeping socially unacceptable or dangerous thoughts at bay. During sleep, he suspected that the censor stopped functioning, letting some of the inappropriate thoughts leak into conscious territory.

In some dreams, the latent content is thoroughly disguised; in others, as in the dreams of children, it's more accessible. Ambiguous latent thoughts are transformed into more legible manifest content through what Freud called "dream-work." Working backward through its component processes — condensation, displacement, considerations of representability, and secondary revision — should enable a psychoanalyst to unravel the dream's meaning and identify the issue at its core.

Through condensation, different elements of the dreamer's life are woven together, subverting the laws of time and space. A character may have the body of one person but the name of another or appear in some incongruous setting. An acquaintance from elementary school might be hanging out at the office; a lecture you'd expect from a parent might be delivered by a public figure. It's one of the most disconcerting characteristics of dreams. As Milan Kundera mused in the novel *Identity,* dreams "impose an unacceptable equivalence among the various periods of the same life, a leveling contemporaneity of everything a person has ever experienced; they discredit the present by denying it its privileged status."

In a related process, displacement, the sleeping mind conflates the important with the irrelevant. Some trivial thing might seem like the main plot; the essence of the dream may appear as a minor detail. Through considerations of representability, thoughts are transformed into pictures and visual symbols. Freud compared this process to the writing of a poem: as the poet creates verses out of feelings and ideas, so the dreamer creates pictures out of latent dream-thoughts. In the final process, secondary revision, the mind gives in to its natural tendency to make order out of disorder and "fills up the gaps in the dream-structure with shreds and patches," uniting the dream's disparate elements into a story with some degree of coherence.

The overwhelming majority of symbols in dreams, according to Freud, refer to sex or anatomy. The list of items that represent the "male organ" include "all elongated objects"; umbrellas (whose opening "might be likened to an erection"); knives, guns, hammers, weapons, and tools; a woman's hat; a man's necktie, because it hangs down his body; and "all complicated machines and appliances." Stand-ins for women and the womb include hollow objects like boxes, chests, cupboards; ships and vessels; rooms, because they have entrances; and tables, "because they have no protruding contours." Climbing up or down a ladder or staircase signifies intercourse. Children symbolize genitals, "since men and women are in the habit of fondly referring to their genitals as little man, little woman, little thing." Playing with a child is, by extension, a symbol for masturbation. Freud insisted to one patient that the color pink, which she associated with the shade of "carnations," actually represented her "carnal" desire.

One patient affected by this new theory of dreams was a Russian aristocrat named Sergei Pankejeff; Freud referred to him as "the Wolf Man." After years of unsuccessful analysis, Freud decided that Pankejeff's depression could be traced to a childhood trauma that he would have forgotten—if he didn't remember the nightmare it had inspired. In the dream, which he'd had around the age of four, Pankejeff noticed a pack of white wolves perching on the branches of a tree outside his window and watched them from his bed, petrified. Freud noted that the wolves were not moving and concluded that his patient had been craving stillness because he had witnessed some frenzied or violent motion at home. He extrapolated, creatively, that the Wolf Man had once walked in on his parents having sex.

The Interpretation of Dreams flopped in 1900; six years after it was printed, only a few hundred copies had been sold. But its reach grew over the next several years, as did the reputation of Freud's psychoanalytic movement, bringing about a new focus on dreams in the profession. One early Freud admirer was the up-and-coming Swiss psychiatrist Carl Jung. In his own life, Jung trusted his dreams, letting them guide important decisions. When he graduated from school, he couldn't make up his mind about what to study next; he was fascinated by science but equally enthralled by history and philosophy. A striking pair of dreams gave him clarity. In one, he was walking along the Rhine River when he came across a grave mound. He stopped to excavate it and was thrilled to discover a pile of prehistoric bones. In the second dream, he stumbled upon a clear pool in a dark forest; when he peered into the water, he saw a shimmering aquatic animal. Jung awoke from these dreams with "an intense desire for knowledge." They confirmed his passion for the natural world, and he embarked happily on his college career in science and medicine.

Jung sent Freud a fawning letter in 1906, and the two men struck up an enthusiastic correspondence. They met in person the next year and talked for nearly thirteen hours. Freud felt that he had finally found a worthy protégé. "If I am Moses, then you are Joshua and will take possession of the promised land of psychiatry," Freud wrote to Jung in 1909. He called Jung his "eldest son," his "successor and crown prince."

But cracks in the relationship soon emerged. Freud felt threatened by the younger man and disapproved of his interest in the supernatural, and Jung came to resent Freud's condescension. One of their most vehement disagreements — one of the issues that led to their final break in 1913 — was about the role of sexuality in the unconscious. Jung criticized Freud for refusing to consider the unconscious as more than a breeding ground for base desires and disputed Freud's obsessive focus on sex in the interpretation of dreams and psychoanalysis as a whole. He agreed that dreams revealed suppressed desires but insisted that those wishes encompassed much more than sexual fantasies. "The dream gives a true picture of the subjective state, while the conscious mind denies that this state exists, or recognizes it only grudgingly," Jung wrote in *Modern Man in Search of a Soul*. "Dreams give information about the secrets of the inner life and reveal to the dreamer hidden factors of his personality. As long as they are undiscovered, they disturb his waking life and betray themselves only in the form of symptoms."

Jung believed that the individual unconscious "rests upon a deeper layer, which does not derive from personal experience and is not a personal acquisition but is inborn." His "collective unconscious" is a basic psychic structure shared by all humankind, built out of a universal set of symbols and instincts that date back to a time before memory or history. It is composed of archetypes, like the Wise Old Man and the Great Mother, that are "found in all times and among all peoples" and are represented in myth, art, religious rituals, psychotic hallucinations, and dreams. The soul is made up of two complementary archetypes, the animus (which represents a woman's masculine energy) and the anima (which represents a man's feminine side). The shadow archetype encompasses the dark, animalistic side of the personality. The most important archetype, the Self, represents the integration of the conscious and the unconscious and different components of the personality. Just as the body regulates itself by maintaining a healthy temperature, so the psyche strives for a balance between the conscious and the unconscious. According to Jung, a man must accept enough of his anima to relate to women; a woman must incorporate her animus without letting it overwhelm her (lest she become too competitive or too interested in life outside the home). The shadow aspects should not be eliminated but integrated.

Jung believed that dreams could help people identify the parts of their psyches that had been overshadowed or neglected. One patient of his, a young man, dreamed that his father was driving recklessly, careening back and forth and finally crashing into a wall. Shocked, the son chastised his father, who only laughed, and the son realized that his father was drunk. In his waking life, the young man idolized his father, a responsible, successful man who would never do anything so dangerous. So why did he cast his father in this uncharacteristic role in his dream? Over the course of their conversations, Jung realized that the young man was overly dependent on his father's approval; his concern about his father's opinions was interfering with his own development. The man's unconscious, Jung decided, was compensating by raising up the son and diminishing the father. This interpretation resonated with the young man, who agreed that he needed to stop giving his father's opinion so much weight.

There was another trend bubbling up in the early twentieth century that helped tilt Westerners' attention toward dreams: anthropologists and ethnologists were taking a new, more serious interest in the culture of their indigenous neighbors. By the 1920s, American Indian arts and crafts were all the rage, and intellectuals were experimenting with hallucinogens in New York salons. And they were fascinated by American Indians' reverence for dreams.

In many indigenous cultures, dreams have been treated as a bridge between this world and the other, a sacred sphere where spirits and ancestors could communicate with the living. A dream about a powerful animal might be interpreted as a sign that a hunt for that animal would be successful. "The Iroquois have, properly speaking, only a single Divinity—the Dream," a Jesuit missionary, Jacques Frémin, wrote in 1669. "To it they render their submission, and follow all its orders with the utmost exactness." (The missionaries may not have grasped the nuances of local belief systems, but their reports enchanted Westerners.) Frémin recalled how one Iroquois man reacted to a dream about bathing: "Upon waking, he ran to several cabins and asked his friends to throw a kettle of water over his body, despite the freezing cold." Another acquaintance traveled five hundred miles to Quebec because he had dreamed of buying a dog there. They felt obliged to act out their dreams, even their nightmares. One missionary claimed to have watched a Huron man chop off his own finger after dreaming that it had been amputated.

Jean de Brébeuf, who spent most of his life trying to convert the Huron to Christianity, also wrote about the locals' respect for dreams. "They have a faith in dreams which surpasses all belief," he wrote. As Brébeuf understood it, they heeded their dreams' decrees, no matter how random or elaborate. When one of his neighbors dreamed of preparing a feast, he ran to Brébeuf's house in the middle of the night, woke him up, and asked to borrow a kettle. If a sick person dreamed that a game of lacrosse could expel the demon that was making him ill, then, "no matter how insignificant the person may be, you will see a fine field where village contends against village for lacrosse supremacy."

When anthropologist Jackson Steward Lincoln embedded with indigenous groups three hundred years later, he found their faith in dreams intact. The Navajo he met explained that shaking hands with a dead person in a dream was akin to a death sentence. (With complicated rituals and good luck, a disaster foretold by a dream might be averted.) The Crow believed that dreams and visions determined an individual's entire path in life; success was credited to good dreams, and failure was blamed on their absence. Crow men would take elaborate steps to invite these all-important dreams; they might isolate themselves in the mountains, haul around buffalo on their backs, or cut off finger joints as a sacrifice.

In one widespread initiation rite, the vision quest, sleep-deprived adolescent boys were sent into the wilderness, alone, to fast and pray. Each boy remained in the woods — often for several nights — until he had a vivid dream about an animal, which granted him secret knowledge and supernatural powers. "The ritual was deliberately designed to place the child in a condition of extreme physical pain and emotional distress — socially isolated, deprived of food and water, exposed to the elements, and vulnerable to attack by wild animals," Kelly Bulkeley wrote in *Dreaming in the World's Religions*. "Judged by contemporary American legal standards, these practices would probably be considered child abuse." By American Indian ones, they were just the opposite; the vision quest was a privilege, an opportunity for a profound religious experience. Its value was not only spiritual but social — surviving the dangerous rite of passage would help the initiate gain status within the tribe.

"The dream is an actual experience, not indistinguishable from waking reality . . . but rather significantly and importantly different from it" was how Michele Stephen, an anthropologist who lived with the Mekeo of Papua New Guinea in the 1970s, summed up her hosts' attitude toward their dreams. For the Mekeo, the dream represented the nighttime action of the soul, which is liberated from the body in sleep. It "allows men access to a realm of knowledge and power usually hidden from them," giving them clues about the future and insight into the secret wishes and intentions of other tribe members. If a sorcerer wanted to speak with a dead relative in a dream, he could take a piece of the body—"usually a finger bone, nail parings, or hair removed from the corpse before burial and kept for this purpose"—and recite an incantation over it. If someone dreamed of being wounded in a fight or attacked by a wild animal, he would avoid public places for weeks. After traveling eighty miles to attend school, one young man—an aspiring teacher—dreamed that he was trapped in a hellscape of fire and evil spirits. He understood his nightmare as a sign that the gods would punish him if he left his family to chase his own ambitions, so he quit the program and moved home. Dreams could cut across the usual social hierarchies: men and women, young and old, were all capable of prophetic dreams that would prompt collective action. If anyone dreamed of catching a fish with a hook, then the entire village must mobilize against the evil spirits on the horizon. As one Mekeo man explained: "The whole village is run by dreams!"

Canadian anthropologist Sylvie Poirier, who lived with indigenous Australians in the Western Desert in the 1980s and 1990s, also marveled at the entanglement of dreams and everyday life. "It was quite frequent for people to share, at will, their dreams, mostly around the campfire where friends and relatives had gathered for morning tea," she wrote. For the Rarámuri of northwestern Mexico, "'What did you dream last night?' is rivaled only by 'How many times did you have sex?' as the most popular morning greeting among men," according to a researcher who embedded with them in the 1970s and 1980s. Dream talk wasn't limited to the morning either. The Rarámuri slept not in one eight-hour stretch but in bouts of a few hours each, giving them ample opportunities to talk about their dreams throughout the night.

• • •

LIKE ANTHROPOLOGISTS AND ethnographers, midcentury psychologists liked to catalog and count, breaking the world into demographically manageable bits. One of the first empirically minded post-Freudian scientists to champion the idea that dreams could reveal buried emotions was a psychologist named Calvin Hall, and he drew on content analysis — a new technique that was catching on in the social sciences — to prove it.

In the 1940s, Hall and a colleague, Robert Van de Castle, began collecting dreams from their students at Case Western Reserve University. As soon as they had gathered a big enough sample, they set to work looking for themes and drawing up norms. They read through each dream as though it were a story, tallying different actions and archetypes, sorting the dream-interactions into categories: failures, successes, aggressions. They counted the number of friends, family members, strangers, and animals. They calculated the ratio of male to female characters. They considered whether the dreamer was making social overtures or only fielding them. They tallied the incidence of activities like eating and sex.

Hall and Van de Castle's analysis revealed a handful of startling patterns. Flying in the face of Freud's hypothesis that dreams fulfill secret wishes, these dream reports were overwhelmingly negative. Aggressive encounters outnumbered friendly ones by two to one; half of men's dreams and one-third of women's featured some kind of physical attack. More than two-thirds of the emotions were negative, with fear, helplessness, and anxiety taking the top slots. There were stark gender differences too, some of which came as less of a surprise. Sex appeared in men's dreams four times as often as in women's. Men's psychic landscapes were dominated by other men — male characters outnumbered female ones by a ratio of two to one — but women dreamed equally of men and women.

Even as Hall and Van de Castle contradicted one aspect of Freud's theory, they confirmed another: dreams could offer an important window into psychic dramas and internal conflicts. When they analyzed people's dreams alongside other measures of their attitudes and personalities, they discovered a noteworthy continuity. Those who were more aggressive in the daytime tended to antagonize others in their dreams, while those who felt powerless were likelier to dream about being persecuted. People with mostly positive interactions in their dreams tended to score high on measures of confidence and social control, while dreams of frustration and anxiety were linked to psychopathology and aggression. All of this might sound obvious, but creating a sort of dream database helped reveal significant exceptions that might otherwise have gone unrecognized. An individual's deviation from the norms could offer clues into how her mind worked — what she thought about, how she related to others, how she saw her place in the world.

A dream world populated by strangers, for instance, is a sign of social alienation; a low friend percentage can be a symptom of mental illness. One study found that friends accounted for only 18 percent of the characters in schizophrenic dreams, and 22 percent in the dreams of the depressed. "If you think about the social situation in which these groups live, it makes perfectly good sense," said psychiatrist Milton Kramer. Schizophrenics dream about strangers, and they "live in a world where their human contacts are decreased. If you count the number of people schizophrenics talk to during the day, it's less than normal people." Depressed people, whose "troubles may be related to their family situation," may dream disproportionately of family structures and family members — if not their own relatives, then people identified by family role: a son, a daughter, a brother.

Thanks to Hall and Van de Castle's quantitative system, psychologists could compare the dreams of different populations no matter how they felt about Freud. Dreaming, they discovered, changed in predictable ways when people underwent episodes of depression or struggled with drinking problems or eating disorders. Tracking a patient's dreams could help them understand dangerous conditions.

As psychiatrists began questioning Freudian theory in the decades after his death, they found that his ideas about dreams didn't always stand up to newer research. Nor did those of his one-time protégé Carl Jung. Jung's beliefs are impossible to test empirically; the existence of the collective unconscious can't be proven or disproven, but it's more plausible that the commonalities in our dreams and symbolic repertoire stem from universal human experiences like growing up, having a body, and learning to participate in a social group. Hall and Van de Castle's discovery that most dreams are unpleasant cast doubt on Freud's idea about wish fulfillment. And the dreams of young children — like Anna Freud and the Wolf Man — figured prominently in Freud's theory, but subsequent studies suggested that children's dreams are typically not sophisticated enough to reflect wishes, hidden or otherwise.

In the 1960s, University of Wyoming psychologist David Foulkes advanced a theory that kids usually didn't remember their dreams before the age of around nine. Even when they were woken up out of a REM phase, children between three and five could recall a dream less than a quarter of the time, and the dreams they did report typically consisted of little more than simple, motionless images. That discovery — like many others in this story — came about not through some clever scheme, but by accident. Foulkes originally wanted to find out if he could manipulate the emotional register of kids' dreams, and he invited children into his lab for a nighttime screening of either a violent or a neutral episode of *Daniel Boone,* a TV show based on an eighteenth-century American pioneer. His results were not particularly interesting; the kids' dreams were no different whether Boone was threatening to "open" captors "from the scalp down" or playing matchmaker for a wealthy miner. But the experiment sparked a more fruitful line of inquiry.

"I had a slowly dawning realization of what a dope I'd been in trying to see how these silly films would change dreams when in fact we didn't even understand what the baseline properties of children's dreams were — no one had done an objective study describing what they were like," Foulkes later told a journalist. Having identified this gap in the literature, he set out to rectify it. He put out ads in the local newspaper and convinced the parents of thirty children between the ages of three and ten to let their kids spend nine nights a year in his sleep lab.

Foulkes woke the children three times each night at strategic points in their REM cycles and asked what, if anything, they had been dreaming about. The youngest kids in the study — the three- and four-year-olds — could report dreams on only about 15 percent of REM awakenings, and the tidbits they did recall were nothing like the complicated stories adults describe. Instead, their dreams resembled fragments or snapshots, images drawn straight from their everyday lives and revolving around basic activities like sleeping and eating. They were brief and unemotional and only rarely included any kind of social exchange. Animals — birds and calves, the kind that populate fairy tales and picture books — figured more prominently than humans. Most revealing was the role the children occupied in their own dreams: they were passive observers, watching the plot unfold rather than directing it or even actively participating. When one of the boys in the study, Dean, was four, he dreamed about sleeping in the bath and napping beside a vending machine.

Among the five- and six-year-olds, rates of dream recall climbed to about 30 percent, and the dream reports grew longer and more complex. Instead of casting animals and food products in the starring roles, school-age children incorporated real people from their lives. When Dean was six, he dreamed of playing games with his friend Freddie at a lakeside cabin, of racing his classmate Johnny on the playground, and of building a Lego bridge at home. It was only around the age of seven or eight that the children finally began to feature as more active protagonists in their own dreams. At about the same time, feelings started to appear in their reports, though the emotional hallmarks of adult dreaming—fear and aggression—were still absent. When Dean was eight, he dreamed that he and five friends planted a seed that sprouted into a tree, withstood a fire, and inspired them to plant a whole forest. In another dream, he found a bunch of balloons in the park, grabbed onto one of their strings, and floated up into the sky.

These trends continued—with recall improving, complexity increasing, and animals receding in importance—as the kids grew up. When Dean's sister, Emily, was twelve, she could report dreams on 86 percent of her REM awakenings, and they were as convoluted as they were common. In one, she found herself on the set of the TV show *Bewitched,* watching fictional characters mingle with her family. In another, she showed her father that she could swallow a piece of her own hair and pull it back out of her mouth.

PANKEJEFF'S DREAM OF wolves began to seem less plausible. (Newer research suggests that kids may remember complex dreams after all, but Foulkes's work went largely unchallenged for decades.) In the 1960s, members of New York's Psychoanalytic Institute spent two years revisiting Freud's theories about dreams and concluded that dream analysis was no longer necessary in their line of work. The generation that came after Freud had developed more sophisticated therapeutic techniques; now, they figured, they could uncover patients' neuroses by talking about everyday activities and waking fantasies.

Not long after, the neuroscientist Allan Hobson cast another shadow on Freud's legacy. Hobson was an unlikely Freud-basher; as an undergraduate, he had been taken with psychoanalytic theory, devouring Freudian texts and even writing his English honors thesis on Freud and Dostoyevsky. But shortly after enrolling in Harvard Medical School—where he decided he would specialize in psychiatry, planning on a career as a clinician—he lost faith in his former beliefs. Freud's ideas didn't jibe with what he was learning about the biology of the brain, and as he absorbed the empirical ethos of Harvard's science department, he grew less tolerant of Freud's haphazard methodology, his preference for his own intuition over data. He was turned off, too, by the "arrogance" of his psychoanalytic teachers, with their obnoxious tendency to analyze everyone in their path, even their own students. At the same time, Hobson gained a new appreciation for the importance of sleep: as a frazzled medical student working around the clock, he caught himself making mindless errors, and he blamed them on chronic sleep deprivation. His dreams became more vivid, as they often do when people sleep in short spurts instead of in one continuous stretch.

In the 1970s, Hobson and another psychiatrist, Robert McCarley, published a novel theory of dreams: they were no more than a reaction to an automatic neural process. Hobson and McCarley implanted microelectrodes in cats' brain stems — the part of the brain then thought to trigger REM — and watched how their nerve cells fired throughout the day. When the cats were awake, their brains were awash in serotonin, which plays an important role in decision-making and learning, and norepinephrine, which helps maintain focus and attention. But when the cats fell into REM, their brains stopped releasing those chemicals and instead secreted acetylcholine, a neurotransmitter involved in emotions and visual imagery. The shifting neurochemical balance caused a flurry of garbled signals to be transmitted from the pons area of the brain stem to the forebrain.

Dreams, Hobson and McCarley extrapolated, were a byproduct of the brain's attempt to make up a story to match this unique combination of chemicals. In their model, it was the physiological state of the brain, rather than repressed memories or deep-seated desires, that determined dream content. Your brain might make you think you're being chased, say, by a knife-wielding ogre not because of your castration anxiety but because your fear centers have been randomly activated. "The forebrain may be making the best of a bad job in producing even partially coherent dream imagery from the relatively noisy signals sent up from the brainstem," Hobson and McCarley wrote. And the reason we forget most of our dreams, according to the activation-synthesis theory, is not that they're too taboo to contemplate — it's that the chemicals necessary for memory formation are missing.

Hobson cultivated his anti-Freud persona, giving numerous interviews and public lectures. He even dramatized his theory in a multimedia exhibit, *Dreamstage,* in Boston — he had a volunteer nap in a glass booth while sleep-lab instruments converted his brain waves and eye movements into blue and green lights that were projected onto a wall in a glowing, dynamic display. The volunteers weren't supposed to sleep at all outside the show so that they would be tired enough to fall asleep on command. That occasionally backfired; one exhausted participant became convinced that Hobson was trying to brainwash her through the EEG machine. Nonetheless, Hobson took *Dreamstage* on tour, teaching around thirty thousand people that dreams were just the product of biology.

While Hobson was busy campaigning against Freudian notions of dreams, an alternative form of therapy, far less concerned with symbols and the unconscious, was taking off. Psychoanalysis was drawn out and expensive, but the new cognitive behavioral therapy was finite, results-oriented, and grounded in scientific research. Whereas traditional shrinks might spend years with their patients, revisiting the wounds of childhood and analyzing their dreams in a potentially endless excavation of the subconscious, practitioners of CBT focused on the present; their goal was simply to help patients overcome depression or neuroses by implementing healthy habits and ditching harmful ones, whatever their origins. And powerful, inexpensive antidepressants and antipsychotic drugs came on the market, making any kind of talk therapy seem dispensable.

By the 1980s, it had become fashionable to denounce Freud's once-venerated ideas as pseudoscience. Feminists attacked him for his penis-centric theories and personal mistreatment of women. Many academic psychologists, eager to be seen as real scientists, didn't want their profession sullied by the controversy. "When the psychoanalytic field splintered, dreams got relegated to the fringe," said psychologist Meg Jay. Discussions of Freud and the unconscious were purged from the psychiatric bible, the *Diagnostic and Statistical Manual of Mental Disorders*. Federal funding for dream research petered out.

It didn't help that the field inaugurated by the Society for Psychical Research back in the 1880s was finally gathering steam. Parapsychology had quietly grown, in fits and starts, ever since the nineteenth century, with institutions sporadically deeming investigations of the paranormal worthy of support. In 1912, Stanford began running lab-based studies of so-called thought transference. A couple of decades later, Duke University opened a parapsychology center, where one-time botanist Joseph Banks Rhine borrowed new techniques from experimental psychology and applied them to the study of psychic phenomena. His colleagues found stories of people who believed they could communicate with each other in their sleep, such as a pair of friends who claimed to have simultaneously dreamed of encountering each other in a burning forest. (The apparent existence of shared dreams was taken as evidence that an astral world existed alongside the physical one.)

In the 1960s and 1970s, just as psychologists were shying away from dreams, the field exploded, benefiting from the climate of cultural openness and spiritual experimentation. Open-minded Westerners did their best to discard the values they had grown up with, dabbling in Eastern philosophy and yoga, Buddhism and meditation. Jon Kabat-Zinn, a microbiologist from New York, set up the Stress Reduction Clinic at the University of Massachusetts in 1979 and taught patients of the hospital's pain clinic about mindfulness and meditation. Inspired by anthropologists' dispatches, Americans began organizing full-scale reenactments of indigenous ceremonies. Seekers tried to purify themselves in sweat lodges — huts heated to mind-altering temperatures. They read about vision quests and set off into the woods to look for their own totem animals. Indigenous Americans were glorified for "their stubborn resistance to white authority, their maintenance of traditional, communal values," wrote historian Philip Jenkins in *Dream Catchers: How Mainstream America Discovered Native Spirituality*. "It is not surprising that a new era of starry-eyed neo-Indianism should mark the decade after 1965, the time of Vietnam and Watergate, of assassinations and urban rioting, of gasoline shortages and threatened ecological catastrophe. . . . Through American history, romantic Indian images are most sought after in eras of alienation and crisis." Tourists to the Southwest purchased indigenous tchotchkes and crafts. Children crafted dream catchers out of feathers and yarn.

Meanwhile, in the academic world, parapsychology labs opened at UCLA, Princeton, and the University of Virginia. The CIA funded secret research into the possibility of weaponizing ESP. In 1962, Montague Ullman, the chairman of the psychiatry department at Brooklyn's prestigious Maimonides Medical Center, convinced his employer to set up a lab for the study of psychic dreams. Ullman made up elaborate rules, determined that skeptics would find nothing to criticize. He screened his volunteers for psychic ability, giving them a test with a 50 percent fail rate. Each aspiring subject would have to spend a preliminary night in the sleep lab while a research assistant stared at one of twelve images and tried to beam it into his mind. Only if the target image ranked in the top six could the participant move on to a real experiment.

Those who passed this hurdle, proving their psychic chops, got to do things like sleep in the lab while a researcher sat in another room, focused on a random painting, and tried to telepathically transmit the image. When machines indicated that the volunteer had fallen into REM, a monitor would rouse him and ask what he had been dreaming about. More than 60 percent of the time, the reported dreams matched the target paintings — according to Ullman's criteria. (A dream about a meal, for instance, could count as a match for a painting of the Last Supper.) In one of their most famous studies, Ullman and his colleagues Stanley Krippner and Charles Honorton asked two thousand concertgoers at a Grateful Dead show to attempt to transmit an image (a man in the lotus pose with bright chakras along his spine) to Malcolm Bessent, a self-styled psychic who was asleep in their lab. Bessent reported a dream about a man "suspended in midair" and mentioned "a spinal column." Ullman counted this a success.

In another study, psychiatrist Robert Van de Castle spent eight nights in Ullman and Krippner's lab. On one night—of which he was particularly proud—a researcher concentrated on a copy of Salvador Dalí's painting *The Discovery of America by Christopher Columbus,* which shows a boyish Columbus stepping onto the new shore, a haloed Virgin Mary clasping her hands on a flag that flies beside him. Van de Castle's dreams that night included "some fairly youngish male figure," "a woman from Atlantic City or Atlantic Beach," and "people dressed in white robes." It was for feats such as this that Ullman and Krippner nicknamed him the "Prince of the Percipients."

"In retrospect, I am surprised that we had as much approval as we did," Krippner admitted recently. "We had no trouble presenting papers at sleep and dream conferences, psychological conferences, et cetera." In the 1970s, the Maimonides team won a grant from the National Institute of Mental Health—a prestigious stamp of approval. "Sixty-five percent of spontaneous ESP experiences take place in dreams," one of the lead researchers told a *New York Times* reporter, who did not challenge him.

In Europe, meanwhile, psychiatrists were attempting to use dreams to predict natural disasters and global events. The first "premonitions bureau" was established in London in response to a disaster in the small Welsh mining village of Aberfan. One Friday morning in October of 1966, a massive mound of coal debris that had built up on a nearby hill collapsed onto a local school, crushing more than one hundred children. As the community fell into mourning and people across the UK learned about the tragedy, some claimed to have sensed something brewing in the days or months before.

As soon as he heard about the accident, English psychiatrist John Barker — who had nursed an interest in paranormal phenomena for years — traveled to Aberfan to find out whether anyone really had predicted the disaster. He put out ads in the national press, and seventy-six people wrote back with stories of their premonitions, about half of which took place in dreams. People from all over England told Barker about their nightmares of children stranded in buildings or killed in avalanches. One woman — the mother of a ten-year-old girl who had died in the landslide — said that the day before the accident, her daughter had insisted that she listen to the nightmare she had just woken from: "I dreamed I went to school and there was no school there. Something black had come down all over it." Another woman said she dreamed that night of a black mass encroaching on the village school. Barker published his results — which many found uncanny — in the *Journal of the Society for Psychical Research,* along with a call to arms: In order to ward off the next disaster, a central office should start systematically collecting dreams and scouring them for warning signs. He and a science writer took it upon themselves to set up the British Premonitions Bureau, and a sister bureau in New York — the Central Premonitions Registry — opened the next year.

Some of the most brilliant minds of the day were swept up in the craze. From mid-October of 1964 until early January of 1965, novelist Vladimir Nabokov carefully tracked his dreams, hoping to prove that they contained hints about the future. His project was inspired by a book called *An Experiment with Time* in which a British engineer argued that time could work backward and that in dreams, we leave the "sleeping body in one universe" and go "wandering off into another," flitting back and forth between the past and the future, revisiting old memories and catching glimpses of things to come. Dreams, according to the book, consist of "images of past experience and images of future experience blended together in approximately equal proportions."

It isn't so surprising that Nabokov would be willing to venture into this outlandish territory; he had never been able to take sleep for granted. All his life, he struggled with a brutal case of insomnia, impervious even to potent sleeping pills. On one night, he recorded in his diary nine separate "toilet interruptions." On another, a red-letter event: "for the first time in years," he had managed to sleep for six hours in a row.

Nabokov had a long-standing habit of transcribing his dreams, and his nights were dizzyingly active, his descriptions characteristically eloquent. "In doomful half-dream saw the scattered streaks of dim light between the slats of the shutters"; "Had rich and strange visions, remembered them between two abysses of sleep." Some of his dreams are classic scenarios; he loses his luggage or misses his train. Others reflect his famous preoccupations with butterflies, literature, and forbidden sex. In an "intensely erotic" dream, he noticed that his own sister was "strangely young and languorous." He drank tea with Tolstoy and chased butterflies with a giant spoon. In a recurrent nightmare, he found himself "in the haunts of interesting butterflies," stranded without his net and grasping at them with his bare hands.

Nabokov didn't have to wait long for what he considered a positive result. On the night of October 17 — just a few days into the experiment — he dreamed that he was meeting with the director of a provincial museum. The two men were chatting beside a tray laden with samples of a rare, precious soil when Nabokov realized — to his horror — that he had been casually nibbling on the specimens throughout their conversation. Three days later, on the morning of October 20, he turned on the television and saw an educational film about the science of soil — geologists were analyzing a tray of samples, arranged in "appetizing little bags." The film, he decided, was the source of the earlier dream. It was, to him, "absolutely clear."

According to British scholar Sue Llewellyn, it isn't completely crazy to think that our dreams are eerily predictive. When we dream, our brains are working fast, processing snippets of information we've picked up and using them to make guesses about the future. "If there are associative patterns in events, they can be used to help predict what will happen next," Llewellyn wrote in the magazine *Aeon*. "Some patterns are deterministic and logical. For example, day follows night. . . . Some patterns are much less obvious. We call them 'probabilistic' because they are based on events that have a *tendency* only to co-occur, so we cannot be as confident in predicting them." When we are awake, Llewellyn explained, "we are good at spotting logical, deterministic patterns," whereas "during REM, we are better at spotting the less obvious or 'remote' associations that predict probabilistic events."

In our sleep, we take stock of vast swaths of information that would be overwhelming to think about consciously. This is the realm of intuition, of knowledge we possess without knowing how we got it. Sometimes, our bodies tip us off: barely perceptible threats come on the horizon, and our hearts beat faster or the hairs on the back of our necks stand up. And sometimes, our gut instincts are dramatized in dreams, converted from mere hunches into vivid stories that demand our attention.

Knowing when to trust them is more an art than a science. For those who can tell the difference between an important dream and a random coincidence, dreams can provide a lifesaving early-warning system. German-born theologian Paul Tillich — who presciently left his home country for America in 1933 — once said that his nightmares helped him recognize the gravity of the political crisis brewing at home. "For months I dreamed about it, literally . . . and awoke with the feeling that our existence was being changed," he said. "In my conscious time I felt that we could escape the worst, but my subconscious knew better." Many ordinary Germans too first sensed the specter of totalitarianism in their dreams. Between 1933, when Hitler became chancellor of Germany, and 1939, when German journalist Charlotte Beradt fled to America, she interviewed more than three hundred of her fellow citizens, including many Jews, about their dreams. To keep her research safe in the increasingly restrictive regime, she invented a secret shorthand — Hitler was Uncle Hans, Goering was Uncle Gustav

— and scattered the coded dream reports, mailing them to different friends in foreign countries. It would be another two decades before she published the results of her study in a forgotten but fascinating book called *The Third Reich of Dreams.*

At the beginning of Beradt's project, when the extent of Hitler's genocidal agenda was not yet known, her interviewees noticed the looming danger in their dreams. Psychologist Bruno Bettelheim noted that many of these dreams were recorded in 1933, just as Hitler was gaining power: "The dreamer seems to anticipate what was going to happen long before it occurred." The dreams revealed latent fears of losing privacy or angering an inscrutable bureaucracy. A middle-aged doctor dreamed that he lay down to relax with a book, only to see the walls of his apartment evaporate. A new sense of terror permeated the everyday, infusing domestic scenes with horror. A housewife dreamed that her oven turned out to be a spying device, and repeated everything she had said in its presence. A greengrocer dreamed that a sofa cushion testified against him. A man dreamed that his radio began blaring "In the name of the Führer, in the name of the Führer" in an endless loop.

EVEN THOUGH DREAMS can help us recognize nascent dangers before they come into full view, dreams that seem precognitive can typically be explained by statistics. We have so many dreams — about four a night — that it's hardly surprising to occasionally spot a similarity between dreams and life. In his book *Paranormality: Why We See What Isn't There,* British psychologist Richard Wiseman estimated that from age fifteen to seventy-five, the average person will have about 87,600 dreams over the course of 21,900 nights. But even people with excellent dream recall forget many of their dreams unless they encounter something in the daytime to jog their memory. So, Wiseman explained, "You have lots of dreams and encounter lots of events. Most of the time the dreams are unrelated to the events, and so you forget about them. However, once in a while one of the dreams will correspond to one of the events. Once this happens, it is suddenly easy to remember the dream and convince yourself that it has magically predicted the future. In reality, it is just the laws of probability at work."

The premonitions bureaus never managed to predict much of anything, and they shut down within a few years. Generations of psychologists have tried to replicate Ullman and Krippner's findings, to little effect. (Krippner, an energetic eighty-five-year-old who travels the world promoting the power of paranormal dreams, blames electrical storms for clogging his successors' psychic airwaves.) "It looked so good, but it has all the problems that free-response research has," said Susan Blackmore, author of several books on consciousness and parapsychology. When people report their own dreams, they generate so much material that it's almost inevitable that an image or two resembles the target. "It's very, very heavily reliant on the method of randomization. And unless you get to the bottom of the randomization, you do not know whether it's been done properly."

Yet the sparse resources for dreams are still sometimes funneled into the investigation of dubious theories. In 1983, European writer Arthur Koestler left his estate to endow a British parapsychology department; the University of Edinburgh's Koestler Parapsychology Unit has been running ever since. In her tenure, the current chair, Caroline Watt, has pivoted the center's focus away from the question of whether or not paranormal phenomena exist and toward the more relevant issue of why so many people stubbornly believe that they do. In one recent study, she found that people were more likely to say that a video resembled their dreams if they already believed in precognition; the belief reinforced itself, priming people to notice similarities between dreams and subsequent real-life events. In another, Watt showed that selective memory — our human tendency to ignore things that contradict our assumptions — helps perpetuate the belief in paranormal dreams; people who read a dream diary and a regular journal by the same person were more likely to remember the dreams that corresponded with real-life events than the ones that didn't.

The psychic studies of dreams were better publicized than they were backed up. The link between pseudoscience, dreams, and outdated psychoanalytic theories took hold of the cultural imagination, making it even harder for those few, renegade researchers to pursue the unpopular science of dreams.

chapter 2

THE VANGUARD

SOME OF THE MOST SIGNIFICANT DISCOVERIES about dreams have been made by outsiders, scientists who had to contend not only with their colleagues' indifference or even outright scorn, but with the practical limitations of working on the margins.

It took the better part of an hour for Eugene Aserinsky to fasten his eight-year-old son, Armond, to the clunky set of electrodes he'd salvaged from the basement of the University of Chicago physiology department. First, he would root around in his son's hair looking for a suitable place to attach the flimsy metal disk. Once he had selected the spot, he would clear it with a razor blade, careful not to take off any more hair than necessary. Next, Armond would hold his breath as his father applied a foul-smelling liquid paste, a conductive material called collodion, to the freshly shaved patch. (Armond still remembers, nearly sixty years later, how the adhesive itched as it dried.) Finally, Eugene would place the electrode in the drop of collodion and wrap it up in tape. "These things were not made for anybody's comfort," Armond said. "All kinds of things had to be rigged to get the damn things to stay in place. Picture the original *Frankenstein* movie."

Eugene was a thirty-year-old graduate student with a spotty resumé and all of his professional hopes staked on this experiment. Though he had always been a talented student, he had bounced around since his teens. He enrolled in local Brooklyn College at just sixteen but couldn't settle on a major; he tried Spanish, social science, and premed but didn't complete the requirements for any of them. Instead, he moved to Maryland and enrolled in dental school. He loved his science classes but soon realized that he hated teeth; his poor eyesight made the delicate work of drilling and sculpting them all but impossible.

He dropped out again and was drafted into the army, where he worked as a high-explosives handler. When World War II ended, he thought he might take advantage of the GI Bill to go back to college. Remembering how much he had enjoyed his biology and physiology lessons in dental school, he applied to the graduate program in physiology at the University of Chicago — a place known to make allowances for promising students with unorthodox backgrounds. "They took him in and said, 'Your academic record is a bit of a Swiss cheese, but you're obviously bright,'" said Armond, now a retired clinical psychologist in Florida. "'Come and let's see what kind of work you do.'"

The whole family would have to scrimp, but Eugene and his wife were eager for him to finally prove himself. He quit his job as a social worker and moved his family into the graduate dorms at Chicago. The Midwestern winters were harsh, and the only source of heat in their apartment was a measly kerosene stove in the living room. "There were constant money worries," Armond remembered. "Many nights I slept with all the winter coats in the house piled on my bed. We didn't have enough blankets to keep me warm."

Aserinsky had hoped to study organ physiology, but he was assigned to work with Nathaniel Kleitman, who presided over the small, unfashionable specialty of sleep. There was "no joy" in their first meeting, Aserinsky later wrote. He was disappointed to be starting his graduate career in a field "dominated by such soft-science types as psychologists." And Kleitman was a forbidding figure; he had established himself as an expert on sleep through sheer force of will. In 1938, he and a colleague spent a month underground in a cave in Kentucky, trying to see if, in the absence of the usual daily fluctuations in light and heat, they could tweak their natural twenty-four-hour sleep cycle. (They tried and failed to switch to a twenty-eight-hour pattern.) Later, he would draft himself into his own study of sleep deprivation and force himself to stay awake for one hundred and eighty hours.

ASERINSKY'S FIRST ASSIGNMENT did nothing to gin up his enthusiasm: Kleitman told his new assistant to stare at babies' eyelids as they fell asleep. He hoped to challenge a paper he had read in *Nature* in which a physicist claimed that he could predict when his fellow passengers on a train would fall asleep by tracking the rate at which they blinked. Kleitman wanted to know whether babies' eyes would stop moving the moment they lost consciousness or if their blinking would gradually taper off. "I was forewarned by Kleitman to bury myself into all the literature on blinking thereby becoming the premiere savant in that narrow field," Aserinsky dryly recalled. After watching babies sleep for weeks, Aserinsky worked up the nerve to go to Kleitman's office and confess that he couldn't distinguish the movement of the babies' eyeballs from the quivering of their eyelids. But he also had an idea: What if he gave up on trying to differentiate true blinking from eyelid fluttering and monitored all of the babies' eye movements as they slept?

Even to Aserinsky, his own proposition "seemed about as exciting as warm milk." But he was allowed to go ahead with it, and after months, he noticed that there were twenty-minute periods in which the babies' eyelids completely stopped moving. Kleitman was intrigued and encouraged his student to keep going, to broaden the project to include adults. "What lay ahead was a gamble—the odds being that since no one had really carefully examined the eyes of an adult throughout a full night's sleep, I would find something," Aserinsky wrote. He hoped he might even be able to use the project as his doctoral research, skipping his bachelor's and master's and finally catching up with his peers. "Of course, the importance of that find would determine whether or not I would win the gamble."

He asked his son to act as a guinea pig, and Armond was excited to help; he didn't mind the tedious setup if he got to spend time with his dad. Eugene would fix the electrodes to Armond's head and turn on the polygraph machine, which translated his brain waves and eye motions into etchings on a continuously rolling sheet of paper.

When Armond fell asleep, the marks appeared on the page as slow, steady ripples; his eyes were still, and his brain, like his body, was dormant. But later in the night, the lines began to oscillate, swinging up and down in jittery waves. It looked more like the polygraph of someone who was awake. He didn't know what to make of this result. Perhaps the old machine was malfunctioning. Perhaps it was just some fluke or an Aserinsky-family quirk. But when he recruited more people to sleep in his lab, he saw the same pattern: four or five times each night, at regular intervals, the brains of his subjects would light up, as active as though they were thinking, talking, walking. And these spurts of cognitive activity coincided with a surge of ocular motion; just when the polygraph indicated that the volunteers' brains were lighting up, their eyeballs would flit rapidly back and forth in their sockets.

Casting around for an explanation, Aserinsky wondered whether his subjects were actually waking up, even though their eyes remained closed. Even scientists like Kleitman — who had devoted his career to sleep — had always assumed that the brain simply turned off at night. Aserinsky waited for his next volunteer to fall asleep and entered the room when his eyes began to move. He tried to talk to him, but the man did not react. "There was no doubt whatsoever that the subject was asleep despite the EEG suggesting a waking state."

Having ruled out the most obvious explanation, Aserinsky let himself entertain a more exciting one. Perhaps those "hoary anecdotal reports" that linked eye movements to dreams might actually be true.

He thought back to Edgar Allan Poe's famous description of the raven: "His eyes have all the seeming of a demon's that is dreaming."

One night, Eugene woke Armond as his eyes darted from side to side and asked him what was going through his mind. "I said, 'You awakened me from a dream,'" Armond recalled. "He asked me for a report of the dream content. It was a fragment, something to do with chickens. He thought that was very interesting. He was very pleased. Talk about great discoveries and humble markers for them."

Aserinsky began prodding his subjects awake at different points in the night and asking whether they could remember any dreams. If he roused a volunteer when the polygraph was quiet and his eyes were still, the person typically had nothing to report. But if he woke a subject during what he had started referring to as "rapid eye movement" sleep, he could usually recall a detailed, story-like dream or two. (He considered calling it "jerky eye movement" but feared inviting taunts "relative to the popular slang meaning of 'jerk.'" "Had I been more courageous," he wrote, "we might be referring today to 'JEM Sleep.'") Once, a sleeping man's eyes began twitching violently back and forth while he called out incomprehensibly and the polygraph went haywire. When the man woke up, he said that he had just been in the throes of a horrible nightmare.

ASERINSKY PUBLISHED HIS results in *Science* in 1953, and his discovery of REM would usher in a new era of sleep and dream research. "I have always felt that this was the breakthrough," sleep scientist William Dement once said. "These eye movements, which had all the attributes of waking eye movements, had absolutely no business appearing in sleep . . . It was this discovery that changed the course of sleep research from a relatively pedestrian inquiry into an intensely exciting endeavor pursued with great determination in laboratories and clinics all over the world."

But the landscape didn't change fast enough to benefit the man himself; the field that Aserinsky's work made possible barely existed when he completed his dissertation. His younger colleague Dement went on to found a famous sleep center at Stanford and become something of a celebrity, with hundreds of publications and even a cameo in the comedy *Sleepwalk with Me*. Meanwhile, Aserinsky, strapped for cash, accepted what Armond remembered as "the first job that came along" — a post at the Bureau of Fisheries in Seattle, examining the impact of electrical currents on salmon — and spent the rest of his career at obscure universities. "He was always disappointed with what happened to his position in the world of sleep."

WHEN STEPHEN LABERGE arrived at Stanford in 1968, scientists were just beginning to accept that dreams were more than a cognitive black hole. But lucid dreaming was a step too far; most doubted that it was even possible. Researchers who had never personally experienced lucid dreams thought the whole thing sounded more like a plot device in a sci-fi thriller than a verifiable phenomenon. How could a person be both conscious and asleep? Philosophers and theologians had made occasional references to lucid dreaming for thousands of years, but perhaps people who believed they were lucid were really briefly waking up; maybe they were lying. But LaBerge knew otherwise.

Born in 1947, the son of an air force officer stationed in Florida, LaBerge was a shy child and took refuge in his own extraordinary imagination. "I was a complete introvert, and had no social skills whatsoever," he once admitted. It didn't help that his family had to pick up and move every few years, in thrall to his father's assignments; by the time he graduated from high school, he had lived in Alabama, Florida, Virginia, Germany, and Japan. He learned to make do with solitary hobbies, entertaining himself by watching movies or tinkering with chemistry sets. One of his favorite places was the local cinema, where he looked forward to catching an installment of the latest serial action film every week. One morning when he was five years old, he awoke from a thrilling dream in which he was swimming through the sea as a kind of amphibian, an "undersea pirate." The dream was so enjoyable that he decided to return to it the next night — and the next, and the next, as though he were going to see the newest episodes of one of his beloved serials. In the midst of one of these dreams, it struck him that he had not drawn breath in quite some time. "I would have the experience of seeing the surface of the ocean far above me and thinking, I can't hold my breath this long!" he reminisced later. "Then I'd think, but in these dreams I can breathe dream-water." Without understanding what he was doing, LaBerge was making his first forays into what would become his life's work. He had figured out how to become conscious within the dream state, how to exert his will over the way his dreams unfolded. It would be almost twenty years before he realized that not everyone's dreams resembled a choose-your-own-adventure story.

In the meantime, LaBerge decided that he wanted to be a scientist. He toyed with chemicals and built his own rockets. "In Germany, for some reason, they didn't mind selling explosives to young Americans," he said. "I made all kinds of explosives." He went on to major in math at the University of Arizona, finishing in just two years. "I was in a big hurry as an undergraduate. *Nothing much to learn,* I thought. *Hurry, hurry, hurry.* Why?" he asked recently, wistful, mocking his younger self. He was eager "to get on with the business of being a scientist," to fulfill the ambition he had cherished since he was a child. He was still a teenager — nineteen years old — when he won a Woodrow Wilson Fellowship to pursue a PhD in chemical physics at Stanford.

He moved to the Bay Area, and there, with the sixties in full swing, he was distracted from his precocious ascent up the academic ladder. California was "ground zero for the hippie movement," he said. "I got interested in the chemistry of the mind." He became fascinated by the problem of consciousness, the mystery of how tiny amounts of chemicals could completely alter his perception and give rise to new worlds in his own mind. He wanted to apply his scientific talent to the study of psychedelic drugs, "but nobody wanted to have anything to do with the topic at the time," he said sadly. "I went to every professor in the chemistry department and asked them: 'I'd like to do work on the chemistry of psychedelics.' Nobody would even think of it. It was just becoming illegal. A tragedy."

So he left Stanford and indulged in "just about everything people were doing those days": Jung, yoga, drugs, transpersonal psychology, "various Buddhisms," meditation. He idolized Bob Dylan and spent a few years teaching himself to play the guitar. "I don't identify with groups very much, but I did identify with being a hippie," LaBerge said. After all of his searching, he thought, *"These are my people."*

He drifted away from the academy, working as a research chemist for a private company and taking his quest farther and farther from the beaten track. In 1972, he found himself at a workshop at the famous New Age Esalen Institute. Tarthang Tulku, the Tibetan Buddhist leading the seminar, could not speak English, but the language barrier didn't stop him from conveying the essence of his philosophy. He stood at the front of the room and repeated two words, over and over: "This. Dream." LaBerge understood: Both dreams and waking reality might be a construction of the mind. In the moment, both were equally valid. That lesson proved a turning point, sparking a new awareness of how dreams might fit into his intellectual journey.

"When I was hitchhiking back to San Francisco, I felt strangely high and expanded just from that exercise," LaBerge remembered. A few days later, he had the first lucid dream of his adult life. He was climbing a colossal Himalayan mountain, fighting his way through perilously high snowdrifts, when he noticed that he was wearing a short-sleeved shirt. "I realized at once that the explanation was that I was dreaming!" he wrote. "I was so delighted that I jumped off the mountain and began to fly away, but the dream faded and I awoke." That brief lucid dream was nothing compared to the spectacular adventures LaBerge would eventually teach himself to induce, but it was enough to reignite his childhood interest in dreams and give him a taste of what he would one day be capable of. He began learning more about Buddhism and Tibetan dream yoga.

In the eleventh century, the Indian Buddhist sage Naropa delineated six yogas for his followers to learn as they moved along the path to enlightenment. Dream yoga was the third, following the yogas of inner heat and illusory body; mastering all six should grant the student access to the bardo state between death and rebirth. In Tibetan Buddhist cosmology, the waking state ranks lowest on the ladder of consciousness; sleeping and dreaming both offer greater possibilities for spiritual growth. The goal of dream yoga is to cultivate an enlightened detachment, to understand that earthly experiences, including dreams, are self-generated illusions. Expert dream yogis can meditate and summon different deities within their dreams.

A few years after his mind-opening experience at Esalen, LaBerge went browsing in the Palo Alto public library and stumbled on a slender book called *Lucid Dreams* by an English scholar, Celia Green. It was a sterile, thoroughly researched volume — a far cry from the mystical sources he had come to rely on. Green had collected case studies through her work at the Institute of Psychophysical Research in Oxford and drawn up basic typologies: actions that were possible within lucid dreams, triggers that could spark lucidity — "emotional stress within the dream," "recognition of incongruity" in the dream world. In measured, academic prose, she laid out some of the characteristics of lucid dreams: "Flying is a common feature of lucid dreams." "Persons who appear in lucid dreams are clearly characterized and retain their identity throughout the dream."

The systematic study of lucid dreams was new, but the idea of lucid dreaming was ancient. In the fourth century BC, Aristotle described the sensation of having a glimmer of awareness while in a dream. "When one is asleep," he wrote in his essay "On Dreams," "there is something in consciousness which declares that what then presents itself is but a dream." In a fifth-century letter to his friend Evodius, the early Christian Augustine of Hippo used lucid dreaming to make the case that consciousness could exist independently of the body and could, by extension, survive after the body gave out. He wrote about the dreams of a physician named Gennadius who had once doubted the reality of the afterlife. One night, Gennadius dreamed of an ethereal young man who guided him to a city of celestial music. It was a vivid, lifelike scene, but in the morning, Gennadius shrugged it off — it was only a dream. The next night, the dream-figure returned and asked whether Gennadius remembered him. He did. "On this the youth inquired whether it was in sleep or when awake that he had seen what he had just narrated. Gennadius answered: 'in sleep.'

"The youth then said: 'You remember it well; it is true that you saw these things in sleep, but I would have you know that even now you are seeing in sleep. . . . Where is your body now?' [Gennadius] answered: 'In my bed.' 'Do you know,' said the youth, 'that the eyes in this body of yours are now bound and closed, and at rest, and that with these eyes you are seeing nothing?' He answered: 'I know it.'" The angelic dream-character then drew a parallel between this dream — with its disparity between Gennadius's subjective experience (arguing with an enigmatic holy man) and his external reality (lying in bed, unconscious) — and the afterlife. "As while you are asleep and lying on your bed these eyes of your body are now unemployed and doing nothing, and yet you have eyes with which you behold me, and enjoy this vision, so, after your death, while your bodily eyes shall be wholly inactive, there shall be in you a life by which you shall still live, and a faculty of perception by which you shall still perceive," he said. "Beware, therefore, after this of harboring doubts as to whether the life of man shall continue after death." Gennadius was convinced.

Intellectual giants of the nineteenth and twentieth centuries wrote about lucid dreaming too. In *The Birth of Tragedy,* Nietzsche described the occasional experience of calling out within a dream, "It is a dream!" and resolving: "I will dream on!" Freud ignored lucid dreaming in the first edition of *The Interpretation of Dreams* but acknowledged in a later version that "there are some people who are quite clearly aware during the night that they are asleep and dreaming and who thus seem to possess the faculty of consciously directing their dreams." Lucid dreaming was finally reified linguistically in 1913, when Dutch psychiatrist and natural lucid dreamer Frederik van Eeden reread his own extensive dream journals and decided that this "particular kind of dream," these dreams that aroused his "keenest interest," deserved their own category. He introduced the term *lucid dream* in a presentation at the Society for Psychical Research, and it stuck. "He who spends a third part of his life in utter unconsciousness better deserves to be called a sleepyhead and dullard," van Eeden wrote in his novel *The Bride of Dreams,* "than he for whom the dark nights are also vivid and rich with pulsing life."

LaBerge was impressed. Here was a body of evidence proving that he wouldn't have to start from scratch. "Among professional sleep and dream researchers, the orthodox view seemed to be that there was something philosophically objectionable about the very notion of lucid dreaming," he later wrote. But Green's book showed that there was precedent for the scientific study of lucid dreams, giving him the courage to stand up to the conventional wisdom. "I was excited to discover that van Eeden was not the only lucid dreamer in Western history." At last, he could envision a way to apply his scientific background to the question that most fascinated him. "There's a group of scientists who study consciousness," he realized. "Dream researchers." He decided to give academia another try, and in 1977, he reapplied to Stanford for a PhD in psychophysiology—a new program bridging psychology and physiology—with a radical proposal: He would study lucid dreaming.

BACK AT STANFORD, LaBerge landed in the same lab as the man who would come to be known as the father of sleep medicine, William Dement. After helping with Aserinsky's early studies, Dement had made an important discovery of his own. The physical movements of the eyes during REM sleep not only signified that the subject was dreaming, but actually corresponded to shifts of gaze within the dream world.

Dement tracked his sleeping subjects' eye movements with an electrooculogram (EOG) and asked them about their dreams when they woke up; he then compared their dream reports with measurements captured by the machines. This system soon bore out his hunch: while the body was paralyzed, the eyes, which were free to move, became a bridge to the outside world. Active dreams tended to produce more marks on the EOG, while inactive dreams led to sparse EOGs. Dement could sometimes even draw a link between specific eye movements and activities in the dream world. Just before one man woke up, his eyes twitched back and forth methodically—left to right, right to left—twenty-six times. When Dement roused him, the groggy volunteer explained that he had been dreaming about a Ping-Pong game; in the moments before waking, he had been eyeing the ball's path back and forth across the table. In another study, a woman dreamed of climbing a flight of five stairs, holding her head high. When she reached the top of the steps, she walked over to a circle of dancers. The polygraph showed a series of five vertical motions, corresponding to her walk up the stairs, followed by a few subtle horizontal movements as she approached the dancers.

Debates about the scanning hypothesis notwithstanding, LaBerge knew of Dement's thesis, and he also knew that, in his own lucid dreams, he could will himself to look wherever he chose. If he could use his eye movements to communicate with a researcher while he was asleep, he reasoned, the reality of lucid dreaming would be impossible to deny. But first, he would have to figure out how to have lucid dreams on command. He would need to find a way to sustain lucidity long enough to signal to another scientist. He would need to not only exert his will over the content of his dreams but extend that control to his physical body. By this point, he was becoming lucid on a fairly regular basis, but he would often wake up as soon as he realized he was dreaming.

The next couple of months were a period of trial and error. All PhD research is intense, but LaBerge's day was only getting started when he left the lab; his real work began when he got into bed. "My doctoral dissertation depended on my doing this," he said. "I had to have lucid dreams in the laboratory. That was a strong motive, but it wasn't enough. I had to have the method." He didn't have one yet. He drew on his years of reading in Tibetan Buddhism, his studies of Celia Green and Frederik van Eeden. Throughout the day, he would think about his goal of becoming conscious in his dreams. When he succeeded, he tried out different methods of prolonging the dream, testing the limits of what he could do, manipulating his dream-environment and moving his actual body — his eyes and even his hands.

The first time LaBerge slept in the lab, he barely remembered his dreams at all. (The only one he could recall was a mundane, non-lucid dream about being in a sleep lab.) Disappointed, he booked himself in for another attempt. The next opening was not for another month — he would have to wait until January, on the night of Friday the thirteenth.

When the inauspicious date finally came, LaBerge had researcher Lynn Nagel hook him up to a polygraph and he climbed into bed. Decades later, he could still remember how everything was set up. "It was a windowless room, the bed against the wall, a little headboard the electrodes go into, and the polygraph a few rooms down the hall. It's dark, completely dark." Seven and a half hours later, he was in the midst of a nondescript, non-lucid dream when he realized how odd it was that he could neither see nor hear. "I recalled with delight that I was sleeping in the laboratory," he later wrote. The next thing he noticed was a pamphlet flying through the void. "The image of what seemed to be the instruction booklet for a vacuum cleaner or some such appliance floated by. It struck me as mere flotsam on the stream of consciousness, but as I focused on it and tried to read the writing, the image gradually stabilized and I had the sensation of opening my (dream) eyes. Then my hands appeared, with the rest of my dream body . . . Since I now had a dream body, I decided to do the eye movements that we had agreed upon as a signal." He used his dream-hand to trace a vertical line and followed the motion with his dream-eyes. Sure enough, a mark appeared on the polygraph. He had done it.

"It's hard to appreciate how strangely wonderful it is," he told me, "to wake up to the fact that you've overcome this barrier of amnesia, that you are communicating with someone in another dimension." He trailed off, a faraway look in his eyes. "Extraordinary."

Over the next weeks and months, LaBerge tried to replicate the feat that he had achieved on January 13, spending night after fruitless night in the lab. Frustrated, he had a polygraph set up at home, and there — from the comfort of his own bed — he managed to repeat the trick another dozen times. He trained three more lucid dreamers — a dancer, a medical resident, and a computer scientist — to send ocular signals from within their dreams. For another project, he fastened electrodes to his own forearms to record the contractions of his muscles. Upon entering a lucid dream, he would clench his fists in a sequence that corresponded to dots and dashes in Morse code. A clench of the left hand would signify a dot; of the right, a dash. In a stunt that should have been impossible to dismiss as mere coincidence, he squeezed out the symbols — left, left, left; left, right, left, left — that represented the letters of his initials: SL.

"It was a huge turnaround," said psychologist Patricia Garfield. "He had physiological records that were taken during the dream state." Before LaBerge came along, "it was thought that lucid dreaming was just drum circles and psychic stuff," said cognitive neuroscientist Erin Wamsley. "Lucid dreaming is a real thing, and LaBerge was the first to really demonstrate that."

Confident in his findings, LaBerge wrote them up and mailed his paper to one top journal after another. "It was something new, genuinely new, in dream research, that was game-changing," he said. "It made it possible to do methodical research." You can almost hear his exasperation in the body of his paper as he anticipated his critics' complaints. "The subjective accounts and physiological measures are in clear agreement," he wrote, "and it would be extremely unparsimonious to suppose that subjects who believed themselves to be asleep while showing physiological indications of sleep were actually awake."

Even so, the paper did not get the reception LaBerge had hoped for. The editor of *Science* thought his findings were too good to be true. It was "difficult to imagine subjects simultaneously both dreaming their dreams and signaling them to others," one reviewer wrote. Others simply ignored him. *Nature* sent the submission straight back, claiming—absurdly—that the topic was "not of sufficient general interest." After six trying months—submitting, revising, resubmitting—LaBerge found a home for his paper in a little-read psychology journal called *Perceptual and Motor Skills*. In the five years following its publication, it was cited by other authors in fewer than a dozen articles.

Meanwhile, LaBerge struggled to move his research forward. Training new lucid dreamers took time, and even competent ones couldn't always perform under pressure. For a few years, he kept doing research in Dement's lab, but the lab was just, "like, half a room in a basement," as LaBerge's wife, Lynne, remembered it. And the lack of funding was a constant burden. "I had to keep finding money from one source or another, when there wasn't any reliable governmental support at all. I found the money by asking . . . individual donors or paying some of it myself." Instead of devoting himself to research, he was forced to hustle. "I had to do things like learn to lecture, learn to teach. As an introvert, all that was pretty hard for me." Dreams were hardly a priority for anyone else. The whole project carried a whiff of the fringe, and the field of sleep medicine—which would one day provide a financial incentive for lucid-dream research—was still getting off the ground.

Even so, LaBerge managed to carry out a handful of intriguing studies. He wanted to strengthen his initial results; even as the evidence piled up, some of his colleagues remained unconvinced. There were "skeptics who said things like, 'Well, there are so many eye movements with REM. Maybe they just happen by chance.'" To LaBerge, that criticism reeked of desperation — decades later, he still got worked up talking about it — but he wanted his proof to be beyond dispute. So, he said, "We tried other channels of communication, to go around that objection." He developed an even more elaborate system of signaling, teaching three lucid dreamers to use their breath to communicate with the waking world; they would hyperventilate and hold their breath in a predetermined pattern to show that they were lucid.

He also found that this system of ocular signaling could help him tackle ancient questions. In one study, he used lucid dreamers to compare the passage of time in waking life and in dreams. He wanted to know how it was possible to live out an entire adventure — skipping across time zones, traveling to other countries and even planets — in the span of a few minutes. Why do people wake from these antics feeling refreshed rather than dizzy and worn out? One explanation was that time worked differently in dreams; perhaps each second of waking time lasted a minute or even an hour in the dream world.

That had been the prevailing theory since the nineteenth century. In 1853, the French doctor Louis Alfred Maury became convinced that all dreams were actually generated at the moment of waking; even scenes that seemed to last for hours, he argued, corresponded to only a few seconds of real time. Maury came to this conclusion after a piece of his headboard fell onto his neck one night, wrenching him out of a circuitous nightmare about the French Revolution: After witnessing a series of murders, Maury saw that it was his turn to die, and ascended the steps to the scaffold. He laid his head on the chopping block, and just as the blade dropped, he woke to find that it was his headboard — not the guillotine — that had landed on his neck. He extrapolated that his brain had made up the entire story — which had seemed, to him, to go on and on — as a near-instantaneous reaction to the collapse of his bed frame.

Now, LaBerge had a new way to approach the question. He asked his subjects to make one signal when they entered a lucid dream and a second when they believed that ten seconds had passed. The second signal arrived, on average, thirteen seconds after the first. Contrary to what Maury would have predicted, the dreamers' perception of time was surprisingly accurate.

In another project, LaBerge explored whether patterns of brain specialization that mark waking activity — with the left hemisphere more involved in logic, and the right in charge of visual and spatial reasoning — persist in the dream state. He arranged to have his own brain monitored as he entered a lucid dream and performed two tasks. One, singing, predominantly engages the right side of the brain when a person is awake, and the other, counting, depends on the left.

When LaBerge became lucid on the night of the experiment, he moved his eyes to signal that he was conscious and launched into a rendition of "Row, row, row your boat." When he finished the song — ending on the apt line "Life is but a dream" — he sent his second signal and counted to ten. Sure enough, the machines showed that the right hemisphere was more active when he sang the nursery rhyme, and the left side of his brain was more engaged when he was counting. "Singing and counting in the lucid dream produced large shifts equivalent to those that occurred during the actual performance of the tasks," he wrote. Just visualizing those activities, though, did not. "This suggests that lucid dreaming (and by extension, dreaming in general) is more like actually doing than like merely imagining."

For another experiment, LaBerge persuaded two volunteers — one male, one female — to incubate lewd lucid dreams in his lab. (A friend of his, psychologist Patricia Garfield, had written rapturously of the "soul- and body-shaking explosions" she enjoyed in her lucid dreams, and he wanted to find out whether lucid-dream sex really could produce the same physical responses as the real-life kind.) The woman, given the pseudonym Miranda, went first. With a probe measuring her vaginal pulse amplitude, she agreed to send a series of ocular signals at key moments: when she first became lucid, when she started having sex, and when she reached orgasm.

On the night in question, Miranda became lucid a few minutes into her fifth REM period. She sent her first signal, flew out through an unopened window, and soared above the arches and carved-stone buildings of the college campus. Shortly after she thought about her erotic mission, a group of men and women materialized. She landed, selected one of the men, and sent her second sign. After about fifteen seconds, she shifted her gaze again, this time to indicate climax. Incredibly, Miranda's physiological measurements told the same story. Between her second and third signals — which bookended the episode of dream-sex — her vaginal blood flow increased, her breathing quickened, and her genital muscles contracted.

LaBerge repeated the experiment on a male lucid dreamer to whom he gave the appropriate pseudonym of Randy. (Instead of the vaginal probe, Randy was hooked up to a penile strain gauge, which measures penile tumescence.) When Randy realized that he was dreaming, he, too, decided to fly, and he floated up through the roof and took off, "Superman-style." He touched down in someone's backyard and, remembering his assignment, "wished for a girl." His wish was granted: a woman appeared and "began to kiss . . . [him] in a most stimulating manner." Once again, the polygraph confirmed parallels between Randy's physiology and the events of the lucid dream. During the thirty-second period of sexual activity, the pace of his breathing accelerated — climbing to its highest level of the night — and his erection strained against the penile gauge. "Remarkably," LaBerge wrote, "a slow detumescence began almost immediately following the dream orgasm." (Unlike in Miranda's case, Randy's dream-orgasm wasn't accompanied by an actual one.)

In spite of his discoveries, LaBerge failed to attract much attention from the scientific establishment. Lucid dreaming didn't seem likely to cure cancer, after all; it was thought of as weird, nonessential, if it was thought of at all. By the end of the decade, the financial situation was dire; in 1988, LaBerge personally owed Stanford twenty thousand dollars. "Had I been able to get funding, I would have stayed solely a researcher, which is really what I am temperamentally suited for," LaBerge once admitted to a reporter.

No one had done more to advance lucid dreaming than Stephen LaBerge. He is to lucid dreaming what Louis Pasteur is to pasteurization, Thomas Edison to electricity. But instead of devoting himself to research, he had to find a way to make money. He set up a private company called the Lucidity Institute and began writing primers on lucid dreaming—like the one I found in Peru.

chapter 3

DREAMS ENTER THE LAB

IF STEPHEN LABERGE STRADDLES THE LINE BETWEEN academic and advocate, driven above all by a passion for dreaming, Matt Wilson is squarely on one side of it. Wilson never intended to make his name in dreams. As a graduate student in computational and neural systems, he was obsessed with the questions of how we form and store memories and how those memories make us who we are; dreams were no subject for a serious researcher like him. But one fateful day in 1991, a rat derailed his plans.

"The thing about natural behavior is that the rat is in control," he reflected from his office at MIT, glancing out of his floor-to-ceiling windows to the sunny Cambridge street below. His office, in the Picower Institute for Learning and Memory, looked like it belonged to a corporate executive rather than an academic scientist; that lab rat launched quite a career. "You try to set a task, but the rats are going to do what they're going to do. After they run, they're tired. It's not like I designed the experiment to make them sleep."

Wilson — then a thirty-year-old postdoc at the University of Arizona — had implanted microelectrodes in his rats' hippocampi and set them loose in a maze strewn with chocolate-flavored treats. He hoped to learn about how the rodents' place cells fired as they ran through the track hunting for the edible rewards. Place cells — a type of neuron that fires whenever an animal arrives in a specific location — play a crucial role in helping animals, including humans, learn their way around new terrain. If, say, a man is foraging in an unfamiliar part of a forest, different place cells in his brain will fire each time he enters a new spot — or place field — generating a mental map of the environment. If he returns to the same area later, his brain can call up the cognitive map he created on his first expedition. The same process takes place when a rat travels through a maze for the first time: new place cells are activated whenever the animal reaches a new spot. The next time it goes down that path, those same neurons fire again, and the rat — whose brain is now familiar with the cells it needs to use — moves more easily through the space. Interestingly, the size of the place field depends on the size of the environment; the longer the maze or the larger the territory, the bigger the place fields. "The first time you're exposed to an environment, the place fields aren't solid," Hannah Wirtshafter, a doctoral researcher in Wilson's lab, explained. "They shift a lot. You don't know what's ahead. As you become more and more familiar with it, they stabilize."

Wilson relied on his electrodes to produce images of the rats' brain waves, but as a backup, he also connected the rodents to an audio monitor. "You can tell a lot about what's going on in the brain by listening to it," he said. "It's a diagnostic for how your recordings are going. You can hear individual cells as they fire. You can hear the different brain states — Is the animal active? Is it running? Is it resting? The different states have different rhythms associated with them. If an animal's active and running, you have what's referred to as a theta rhythm, a ten-hertz rhythm."

That day in 1991, he wrapped up a round of the experiment, took the rat out of the maze, and dropped it back into its cage. Tired and satiated, with nothing left to chase, the rodent promptly fell asleep. But Wilson's job was far from over, and he began the chore of processing his new data. He happened to leave the audio monitor plugged in and was hard at work when he noticed something odd. "Suddenly I heard activity that sounded like the animal running. I heard this theta rhythm — 'ch-ch-ch.' And then I heard place cells — 'bup-bup-bup.'" Wilson was confused. He didn't think the rodent was moving anymore; he had just watched it fall asleep. Why was he hearing that distinctive ten-hertz rhythm, the one associated with running?

"At first this was concerning to me," Wilson said. "I thought the animal must have gotten up — maybe it was going to jump off. I turned to see what was going on, and the animal was still asleep." Stunned, Wilson dropped his data and homed in on the sounds streaming through the speakers. There was no mistaking it: the animal's place cells were firing in the same pattern he had heard as it traveled through the maze. "It just jumped out," he said. "The cells were firing as though the animal were running, but the animal was sleeping." Its neurons seemed to be replaying the task it had performed while awake. "There was spillover of activity from wakefulness," Wilson said. "I could literally hear the brain dreaming."

Realizing he had a major discovery on his hands, Wilson turned his focus away from rats' daytime cognition and began studying the nocturnal rodent brain instead. His new direction was rewarded with one groundbreaking result after another. He figured out how to track the rats' cellular firing so precisely that he could tell which part of the maze it was replaying in its sleep. He learned that the more often an animal visited a certain area in the day, the more often the associated place cells would be reactivated in sleep, as though the rat's brain was reserving its sleeping energy for the most important events of the day. He noticed that the cellular repetition of daytime activity was not confined to the hippocampus; sensory areas of the brain, like the visual cortex, would also light up as the rats slept, suggesting that they could actually be experiencing visual imagery in their sleep.

Could he really call this neuronal replay *dreaming*? We can't ask a rat about its subjective experience, but Wilson thinks it's a fair assumption. "Real memories that change over time and have imagery associated with them — I would consider that to be a reasonable definition of what a dream state would look like, at least in a rat," he said.

When Wilson published the results of his rodent research in the 1990s and early 2000s, he helped revive the scientific study of dreams. Skeptics could dismiss research that relied on human self-reporting or that came from a messenger who didn't fit in, but it was impossible to deny the veracity of Wilson's work. Rats, unlike people, can't lie about dreams that deal with taboos. Nor does it matter if the rats can "remember" their dreams after waking. And rat brains provide "an accessible model," Wilson pointed out. "Being able to study dreams in a controlled system, having access to an animal model of dreams, has been hugely influential in changing the perspective on sleep and dreams." The activity of their neurons can be observed without too much trouble. Their daytime environment can be controlled. And while humans' cognitive systems are exponentially more complex than rodents', it's still possible to draw useful parallels between the two; the most relevant brain structures fill similar functions in both species. "The hippocampus is involved in space in humans as well as in rodents," Wilson explained. The hippocampus — a seahorse-shaped area deep inside the brain and the host of many of those place cells — plays a key role in every phase of spatial learning, from encoding new memories to consolidating and recalling them. London taxi drivers famously develop enlarged hippocampi as they learn the thousands of streets and landmarks of their sprawling city. On the flip side, hippocampal damage — in rodents as well as humans — can render spatial-memory tasks all but impossible.

WHILE WILSON WAS listening in on rats' dreams at MIT, another seminal study on sleep and learning was under way just a few miles down the road. At the turn of the millennium, Harvard psychiatrist Robert Stickgold published a game-changing discovery of his own: If students played Tetris during the day, images of the game would turn up in their dreams that night. It looked as if they were continuing to practice in their sleep.

Stickgold can trace his hunch that dreams — specifically, sleep-onset dreams, or hypnagogia — encode our most potent memories back to an experience he had while rock climbing in the wilderness. "I had been staying up in Vermont with my family," he recalled, speaking in his unassuming office on the eighth floor of a Boston hospital. The plaque outside his door read BOISTEROUS BOB, and it was instantly clear that he deserved the nickname; his excitement over long-ago discoveries, his glee at the serendipity of the scientific process and the cleverness of his students, were palpable even as he reminisced so many years later.

At the end of one grueling day scaling a mountain called Camel's Hump, Stickgold crawled into bed, exhausted and sore, looking forward to a refreshing night's rest. But just as he was beginning to doze off, he was struck by the illusion that he had been transported back to the site of that day's adventure. "As I was falling asleep, I noticed that I was vividly replaying the memories of being on the mountain," he said. There he was again, struggling to pull himself out of a particularly difficult passing. "I could feel rocks under my fingers." He wrenched himself out of his reverie, but as soon as he started to fall back to sleep, it happened again; he couldn't shake the feeling that he was still on the rocks. "It was already six or eight hours after we were off the mountain. It looked like sleep itself was doing something to those memories."

For the rest of his vacation, Stickgold tuned in to the images that rose up in his mind as he fell asleep. He noticed that they often featured challenging activities he had recently attempted — sailing through stormy waters, whitewater rafting — and he began to suspect that his brain was replaying the most salient or stressful events of the day.

When he got back to Harvard, he wanted to test his theory, but he couldn't imagine a way to do it in the lab. "I laughed and said, 'So now I just have to put in an IRB proposal to take a bunch of subjects mountain-climbing,'" he said. One day, he was complaining to a group of students about the impracticality of studying the phenomenon he had noticed in Vermont. "And one of them said, 'You know, the same thing happens with Tetris.'" On the nights after playing the computer game — manipulating colorful shapes as they fell to the bottom of the screen — the student would often see geometric tiles in his dreams. Thanks to that conversation, Stickgold said, "I realized that there was a format we could study it in."

He rounded up twenty-seven subjects: ten "experts," who had already logged at least fifty hours playing Tetris and seventeen "amateurs," who had never played before. Stickgold had all the participants practice Tetris for seven hours over the course of three days. Several times during the subjects' first hour of sleep, Stickgold or one of his assistants would wake them and ask what they were dreaming about. More than three-fifths reported images of falling bricks, just like the virtual ones they had arranged during the day. The Tetris-themed dreams were most common among the amateur players; three-quarters of the novices dreamed about the game at least once, compared to about half of their more experienced peers. Just as Stickgold's drowsy mind had taken him back to the trail in Vermont — giving him another shot at a difficult climb — the subjects' sleeping brains were giving them an extra chance to rehearse their new skill.

The amateur and advanced players also incorporated Tetris into their dreams in different ways. The novices dreamed about black-and-white shapes just like the ones they had seen on the computer, whereas the veteran players' dreams sometimes included more loosely related memories. One woman dreamed of colorful tiles that reminded her of the model of Tetris she had learned years earlier. Her brain, Stickgold said, was revisiting old experiences "to alter their strengths, structures, or associations in ways that are adaptive." The dreaming mind can call up distant, even forgotten, memories in the service of helping us master similar tasks.

Just for the sake of it — he chalks it up to a fortuitous friendship with a local psychiatrist, who was working with amnesia patients that summer — Stickgold threw in, among the novice players, five who suffered from amnesia; lesions in their hippocampi had left them incapable of forming or retaining new memories. David Roddenberry, Stickgold's undergraduate research assistant, commuted to the amnesiacs' homes in the suburbs of Boston. Every time he visited, they would say things like, "Oh, hello, have we met before?" Roddenberry would patiently reintroduce himself and explain — whether for the first time or the third — the rules of Tetris. "They were just really nice," he remembered. "They didn't have any negative associations or distrust," since they forgot about arguments and slights before they could mature into grudges. At each patient's bedtime, Roddenberry would hook the subject up to a monitoring device and take up his post in the next room. When a computer alerted him that the subject had entered REM, Roddenberry would tiptoe into the bedroom, tap him on the shoulder, and ask what was going through his mind. "They would be startled, because they didn't know who I was or why I was in their bedroom," Roddenberry said.

But the results of the study made those awkward moments worth it. Although the amnesiacs couldn't retain even basic memories of Tetris from day to day, they still saw images of bricks in their dreams. "They would describe blocks floating, or they would be trying to line things up but wouldn't know what they were trying to line up," Roddenberry said. They couldn't consciously connect the images with the computer game, but their dreams had clearly been influenced by their play.

"I remember the first time David called me and said, 'Bob, we've got a report,'" Stickgold said. "I was truly stunned. I remember walking out of my office into the lab and saying to myself, Why were we even trying this? It was outrageous to imagine it would work." His random decision to include amnesiac patients ended up yielding the biggest surprise of the entire project. "In an ironic way, it was the proof of Freud's theory that dreams are the royal path to the unconscious," Stickgold said. (Though an action-figure Sigmund Freud is affixed to his bulletin board, he identifies as strictly anti-Freudian.) "These are memories that the amnesiacs clearly had but didn't have access to."

Those Tetris dreams could also help explain why the amnesiacs' scores improved slightly over the course of the three-day study in spite of their inability to remember how the game worked. Toward the end of the experiment, one amnesiac woman sat down at the computer and spontaneously arranged her fingers on the three keys she would need to use.

In 2000, "Replaying the Game: Hypnagogic Images in Normals and Amnesics" became the first paper on the topic of dreaming to make it into the hallowed journal *Science* in more than thirty years. "The Tetris paper was big because it said, 'You can do science. You can use this paradigm and start to ask questions about what gets into your dreams, what doesn't get into your dreams, how it gets into your dreams.'" It paved the way for other scientists to explore the role of dreams in learning—which would eventually be recognized as one of their most crucial functions.

MEANWHILE, IN CANADA, psychologist Joseph De Koninck was investigating the relationship between dreams and language acquisition. De Koninck dates his interest in that topic back to his own young adulthood. Born to Anglophone parents but raised in French-speaking Quebec City, he didn't have to conduct his life in English until moving to Manitoba for his PhD. Though he had a solid grasp of the language, he struggled with the nuances, and his work suffered. "I was always translating calculations from English to French, making the calculations in French, and going back and forth," he said. He carried on that way for weeks until one day, something clicked, and it all fell into place. Suddenly, he said, "I was able to think in English." Around the same time, he noticed something else. "I was starting to dream in English as well as in French."

He wondered whether there might be a relationship between the shift in his dreaming and his breakthrough in waking life, and it was one of the first questions he tackled once he had his own lab at the University of Ottawa. He began by recruiting groups of Anglophone undergraduates who were working to polish their French in a six-week crash course. Although they had enrolled in a bilingual college, their French skills hovered around the high-school level, and they had forfeited their summer holiday to bring them up to par. They attended language classes all day and immersion activities at night and lived together on campus, practicing French around the clock with their fellow classmates.

The students agreed to sleep in De Koninck's lab and tell him about their dreams at three critical junctures: before the program began, a few weeks into it, and after it ended.

De Koninck's discoveries matched his own experience. At first, French was largely absent from the students' dream reports. "Most of your dreams are related to activities you've done the day before, but even if you spend a whole day in immersion, you don't immediately incorporate French in your dreams," he explained. "It's difficult to dream in a second language if you don't master at least some of the grammar, because you can't produce it in your dreaming." After a few weeks of near-constant immersion, though, snippets of French began trickling into the dreams of the best students. "Those that started to incorporate French in their dreams were those who were learning the language more quickly, or had achieved 'mastery,'" De Koninck said. "It does come in, but it takes a while."

When he compared the students' scores on French tests with data on their sleep cycles, De Koninck found another striking pattern. Those who spent a higher proportion of the night in REM sleep tended to make better progress; more time in dreaming sleep went hand in hand with mastery. In fact, except for the three students whose French didn't improve at all, everyone in the program spent more of the night in REM sleep during the intensive course than before or after it. Their brains went into overdrive, giving them extra opportunities to dream about — and strengthen — the new skill.

A few years later, De Koninck designed a more dramatic study on the link between dreaming and mastering a new mindset. He had his students wear glasses that inverted their visual fields, making everything appear to be upside down. At first, they struggled with even basic activities. "They had to relearn to read, to walk," De Koninck remembered. "They were young and they thought it was funny, but it was quite demanding. I don't know if we could still do that today."

This was a round-the-clock ordeal; the students didn't even get a break when they got into bed. On the nights after they had worn the goggles, half of them dreamed of upside-down people or objects. Even those who didn't dream directly about inversion were processing their altered reality in more oblique ways, dreaming of falling down or of feeling confused. "The observed changes in dreams," De Koninck wrote, ". . . reflect the waking preoccupation and psychological state associated with visual inversion." The students also had a "tremendous" increase in REM sleep — a jump that was especially dramatic among the ones who successfully adapted, who stopped fumbling and figured out how to read, sort cards, and even copy text upside down. Whether they were benefiting in real life from their unconscious practice sessions or whether the inversion dreams were only a reflection of the work they were doing in the day, there was a link between dreaming about the task and actually excelling at it.

WHEN I VISITED Wilson's lab at MIT, the place was buzzier than I had imagined it would be; the institute was state-of-the-art, the graduate students cheerful. Sleep science was clearly a priority here.

And the pace of the experiments that were under way helped me appreciate just how exciting it must have been when Wilson heard that sleeping rat's neurons light up all those years ago; the normal pace of science is so much more tedious. I spent a lot of my visit making small talk with graduate student Hannah Wirtshafter as we waited hours for a rat to fall asleep. The rat, whose circadian rhythm had dictated my day and Hannah's winter, sported a bald patch on its head where Hannah had shaved him before implanting thirty-two tetrodes in his brain in what she remembered, with a shudder, as "a horrible six-hour surgery." I was struck by the amount of care and attention paid to a single rodent. Hannah already knew the rat's idiosyncrasies, although it had been recovering from anesthesia for all four days of their acquaintance. It was a fast learner. Sometimes, it would use its tail as a sleep mask, wrapping it around its face to shield its eyes from the electronic glow of the many devices set up to record the firing of its cells. One wall of the room was lined with screens displaying real-time maps of the electrical activity of the rat's neurons. Along another wall was a mass of cords and aluminum foil, a sound system worthy of a DJ booth, amplifying the noise of the rat's cells as they fired. As Hannah explained her project, the rat's neurons hummed continuously in the background, buzzing in a kind of white noise. Occasionally, Hannah cut off the conversation and tuned in to the vibrations coming through the speakers or turned to check one of the screens in front of us.

"There!" she would exclaim as the rhythm changed in a manner imperceptible to me. She pointed to the rainbow waves on the monitor. "We're hearing theta right now." Hannah knew, without looking at the rat, that it had gotten up, that it was running around the lid. Later: "Did you hear that *bum-bum*?" (I did not.) "That's not a neuron — that's him scratching or chewing. You learn to always be listening to it."

In the 1990s, scientists like Wilson, Stickgold, and De Koninck showed not only that dreams help us learn, but that dreams could be studied in the lab. Pedigreed, presentable academics, they were ideal ambassadors for that message. Even so, the patience of scientists who want to study dreams is sometimes tested.

chapter 4

THE RENAISSANCE OF SLEEP RESEARCH

"I DON'T SAY SO MUCH TO MY COLLEAGUES, THAT I come to this conference," Marc, a silver fox of a Belgian psychiatrist, said quietly. He leaned in. We were whispering over our plates of stodgy cafeteria conference-fare. "They would think that I had lost my mind."

"My brother tells people I'm going to a psychology conference," said Ange, a librarian from Toronto.

She wasn't — not exactly. Marc, Ange, and I — and about three hundred others — had traveled to a far corner of the Netherlands and gathered in a medieval abbey, whose grand corridors were decked out with psychedelic collages and distorted faces fashioned out of clay, works of art inspired by their creators' dreams. The status of dreams within the sciences has waxed and waned, but for thirty-five years — ever since a handful of psychics and scientists, tired of being treated as second-class citizens at sleep conferences, joined forces to fight for the unusual passion they shared — the annual conference of the International Association for the Study of Dreams (IASD) has been a refuge for dream scholars and amateurs alike.

When I first heard about the existence of such a thing, I thought my project might end there: Was this going to be a one-stop shop for my questions about dreams? A gathering of the brightest minds in different fields — neuroscience and psychology, history and literature — coming together to get to the bottom of why we dream and what dreams mean? By the time I had done some research and gone to check it out for myself, though, I had no idea what to expect. The website's recommended reading listed anthropological monographs and scientific case studies as well as self-help on "how to dream your future." The organization's board included academic psychologists and a biochemist but also self-styled personal coaches who believed that dreams should determine every decision in life. Browsing the program for the conference didn't help either. There would be lectures on the cognitive neuroscience of lucid dreaming and on the use of "energy fields" in dream interpretation.

Like any subculture or scene, IASD has its own conventions and norms. By day three, I hardly blinked when a conversation started with an inquiry into my lucid-dream frequency or the quality of the previous night's dreams. "Do you work with your dreams?" "How's your dream life?" I was asked again and again, as I might be asked about my job at a cocktail party in New York or about my hobbies on a bad date. For many of these dream enthusiasts, the annual conference of the IASD is the high point of the year. "When I got here, I felt that I had come home," said Sherry, a woman who travels the United States teaching classes on the power of dreams. "It's being part of a tribe," said Walter, a lighting designer who had flown in from California. "It's a place I live for."

I couldn't tell just from looking who belonged on which side of the aisle. When I first caught sight of Harvard psychiatrist Deirdre Barrett — who has edited four academic books about dreams — she was roaming the abbey gardens in a floor-skimming skirt and a T-shirt printed with a quote from Slavoj Zizek: "Reality is for those who cannot endure their dreams." David Saunders, a young British academic, tied up his waist-length hair in a black bandanna and favored black suits on even the hottest days. A black bracelet emblazoned with the question AM I DREAMING? was always around his wrist. His dissertation is one of the most thorough investigations of whether everyone can learn to lucid dream, and he listened patiently as another conference-goer explained the lucid-dream-induction device he was building in his basement.

The members of IASD revel in the organization's strange history, upholding traditions like the Dream Telepathy Contest. One presenter at the conference, nominated as the sender, privately selects an image to meditate on; other guests drop off dream reports the next morning, and whoever's dream most resembles the sender's target image is declared the winner.

On one of the last nights, a lucid dreamer named Clare, who had been appointed that year's sender, retired to her room and shouted into the darkness, "Elephant-headed god!" She flapped her arms and pretended to be an elephant. She went to sleep and fell into a lucid dream. In the lucid state, she flew to the clock tower and beamed an image of the elephant-headed Hindu god Ganesha onto the abbey's deserted grounds. That night, Loren, a poet who had traveled from Seoul, dreamed of an animate elephant-shaped penholder with black-and-white pens for tusks. In the morning, he wrote down all of the details he could remember and checked the pictures a contest organizer had hung in the corridor overnight: a pole vaulter, a baby in front of a Christmas tree, horses in a meadow, a delicately drawn Ganesha. He dropped his dream into the box beside the last picture. The next day, Clare congratulated him and ceremoniously placed a yellow paper crown on his head.

On the final evening, we gathered in a nondescript conference room for the long-awaited Dream Ball: the climax of the week, where presenters and attendees alike dressed up as characters and acted out scenes from their own dreams. In an all-too-literal reenactment of her nightmare, according to IASD lore, a woman once turned up naked.

This year, a woman in a strappy purple dress danced provocatively with a straw hat. A woman wearing blue fairy wings and a wreath of flowers in her hair broke into song: "'I walked with you once upon a dream . . .'" A woman draped in strings of Christmas lights tiptoed in a circle, dropping a trail of red feathers in her wake. A man in a floor-length cape and black feather mask came as his Jungian shadow self. I lost track; they started to blur into an indeterminate mass of glitter and sequins, bubble wrap and wigs. Someone was a mirrored room. Someone was a witch; the sky; the color white.

They formed a line and explained their costumes in monologues that ended in epiphanies. "And then I realized, I am on my path in life." "And then I understood the true nature of reality." It felt like a cross between a group therapy session and a school dance — if the teachers were out on the dance floor. Neuroscientist Michael Schredl, who runs one of the most productive sleep labs in Europe, sang along as a man in a synthetic blond wig strummed a guitar. Barrett brought an owl mask.

"A lot of people feel grief because they go home and don't have people to share their dreams with," said the president of IASD, who wrote a book about karmic healing. I felt like I'd intruded on a secret meeting of missionaries as I listened to the leaders swap ideas about how to make the rest of the world more dream-friendly. One suggested we advertise on our cars and our bodies, issuing indiscriminate invitations on bumper stickers and T-shirts: TELL ME YOUR DREAMS.

THOSE CAMPAIGNS MIGHT soon sound less far-fetched. Dreams and sleep are emerging from the shadows at last. Today, the scientists who study the mind at night have chairs in departments of medicine and neuroscience. They have MRI machines and modern labs; they no longer need to build their own tools or convene in distant monasteries. When they dress up for the Dream Ball and toast the most telepathic dreamer, it's in homage to their predecessors on the fringes, to the ideas that couldn't be proven. And it's a testament to the intellectual open-mindedness that has always set dream researchers apart, from Eugene Aserinsky and Stephen LaBerge to Deirdre Barrett and Robert

Stickgold. Maybe the scholars drawn to dreams are already unusually tolerant; maybe their immersion in the world of dreams makes them more so. I have sometimes found this project destabilizing; obsessing over the mysteries of dreams can leave me feeling unmoored, disconnected from the physical world, more prone to lapses in logic, more skeptical of any certainty.

THE REVIVAL OF scientific interest in dreams — the thrust that has rendered IASD a novelty rather than a necessity — stems from our newfound appreciation for sleep. Over the past couple of decades, insufficient sleep has been linked to a host of physical and psychological issues, from poor focus and impaired problem-solving to anxiety, depression, heart disease, and weight gain. In 2000, a pair of psychologists compared the effects of sleep deprivation and alcohol intoxication and showed that going without sleep for seventeen to nineteen hours led to the same level of impairment in hand-eye coordination, memory, and logical reasoning as having a blood alcohol concentration of 0.05 percent (about two drinks). Drowsy driving is now recognized as a hazard along the lines of drunk driving. "Every hour," according to sleep scientist Matthew Walker, "someone dies in a traffic accident in the US due to a fatigue-related error." In a 2002 experiment, half of a group of medical residents at a Boston hospital were given a reduced workload and asked to track their work and sleep patterns. On the traditional schedule, residents worked on average nineteen more hours per week and slept about forty-nine minutes less per night — and they also made more than twice as many attentional failures (lapses in focus marked by slow-rolling eye movements) during evenings at work. Follow-up studies of doctors and residents revealed a relationship between sleep deprivation and serious — even fatal — errors in diagnosis and prescription.

Sleep, scientists have learned, is the most important phase in the body's cycle of cellular repair. The glymphatic system — which helps the brain flush away toxic cellular waste — accelerates. In mice, myelin

— a fatty substance that protects nerve fibers and facilitates communication between neurons — regenerates. Human growth hormone — involved in growth in children and various metabolic processes in adults — pours out of the pituitary gland.

Sleep loss elevates the risk of heart attack and stroke and can weaken the immune system. Chronic sleep deprivation is a risk factor for hypertension, and even a single all-nighter can trigger an unhealthy rise in blood pressure. Poor sleep sends the appetite haywire, spiking levels of ghrelin, a hormone that stimulates hunger, and slashing stores of leptin, which suppresses it. One longitudinal study that followed nearly five hundred adults over thirteen years found — after controlling for variables like age, physical fitness, and family history — that young adults who slept less than six hours a night had obesity rates more than seven times as high as their peers. Another study of about fifteen hundred adults found that people who regularly slept for less than five hours were more than twice as likely to be diagnosed with diabetes as those who logged seven to eight hours a night.

Research has shown that sleep plays an irreplaceable role in maintaining mental health and helping us process painful memories. People who are forced to look at upsetting images are less disturbed by a second viewing if they've slept in between. And just a single night of poor sleep can create temporary problems — irritability, paranoia, anger — in even the most psychologically sound. "My body had no more feeling than a drowned corpse," said a once-lively woman in a Haruki Murakami story after a spell of insomnia. "My very existence, my life in the world, seemed like a hallucination."

Historically, our cavalier attitude toward sleep has allowed for some risky — but ultimately enlightening — stunts. In January of 1959, a popular DJ named Peter Tripp decided to raise money for charity by staying awake for two hundred hours — more than eight days. A healthy thirty-two-year-old, he figured he would be fine. He arranged to live in a glass box in Times Square from which he could entertain tourists and broadcast his nightly radio program. He would also pay regular visits to the nearby Astor Hotel, where a team of psychiatrists had set up shop. Doctors monitored him around the clock, taking it in turns to make sure he didn't catch even a moment of sleep. When Tripp started to flag, they would tell him jokes, play games with him, or shake him back to life.

The *New York Times* kept the public up to date with near-daily dispatches on his progress. On the first day, Tripp was his usual upbeat self—laughing and waving through the glass—but his mood was the first casualty of his ordeal. By day three, he was snapping at everyone around him. He berated his barber so vehemently that the man broke down in tears. He managed to pull himself together every evening, though, spinning his records and chatting with listeners on his radio show. At the halfway point—when he had been up for one hundred hours—the *Times* reported that he appeared "tired but normal in physiology" and "capable of carrying on an apparently normal conversation."

Tripp's grip on reality went next. He took off his shoes and wondered why no else cared that an army of spiders was parading across the soles. He became convinced that a clock in Times Square was actually his friend's face. His articulate, radio-caliber speech devolved into a mumble, and his body temperature dipped to a dangerous low. After Tripp had been up for a hundred and thirty-five hours, the psychiatrists' wit was no longer enough to keep him awake, and they began feeding him Ritalin as often as four times a day. The stimulant kept him going, but it couldn't restore his sanity. Toward the end of the experiment, he ran away from one of his doctors, thinking that the man, who was dressed in a suit, was a funeral director coming for his body. "On the very last night, he was so far out of it that we were hard-pressed to keep from calling an end to the whole thing," one of the researchers later confessed. After two hundred and one hours—nearly eight and a half days—Tripp stumbled to the Astor Hotel and slept for thirteen hours and thirteen minutes, spending the better part of that time in REM. The next morning, his hallucinations ceased and his mood bounced back, but the researchers involved have since admitted how risky the project was. "We did it carefully, but I think if we had thought about it a little more carefully, we would have had some objections to it," said one of the lead scientists.

A few years later, in 1965, Randy Gardner, a seventeen-year-old high-school student in San Diego, was casting around for a science project when he heard about Tripp's feat. With the bravado of a teenage boy, Gardner figured he could do it better and resolved to break Tripp's record by staying up for 264 hours — eleven days. William Dement read about Gardner's plan in a local paper and volunteered to help keep him awake — a task that became more and more demanding, especially between the hours of three and seven in the morning. When Randy begged to be allowed to close his eyes, Dement yelled at him or sent him outside to play ball. He managed to stop Randy from sleeping, but the nightly vigils took a toll; at one point, Dement was pulled over for driving the wrong direction down a one-way street.

Meanwhile, Gardner's physical and psychological health fell apart. By day two, he struggled to focus his eyes and had to give up one of his favorite distractions, TV. On the third day, he complained of feeling nauseated and started to lose control of his muscles. The following day, his mood darkened, and he felt as though a tight band had been wrapped around his skull. On the eleventh day, he was asked to count backward by sevens starting with one hundred. He got down to sixty-five and abruptly stopped; when asked why, he said that he couldn't remember what he was doing. When Gardner turned up at the hospital 252 hours in, his condition was altogether alarming. He suffered, according to the doctor's notes, from "intermittent irritability, incoordination, slurred speech, partial nominal aphasia, difficulty focusing his eyes," and delusions. His voice was "soft, slow, slurred, and listless, with no inflections." His face was expressionless. His eyelids drooped, and his arms twitched.

In 1989, the *Guinness Book of World Records* deemed sleep deprivation too dangerous and deleted the category.

Sleep, we now know, is crucial for learning. Lapses in memory are one of the most obvious effects of sleep deprivation, as anyone who has pulled an all-nighter knows. Even a single night of insufficient sleep can wreck our ability to learn new skills and absorb new information. In the lab, well-rested volunteers outperform sleep-deprived peers at all kinds of memory, spatial, and cognitive tests — recalling details or words they encountered the night before, navigating a virtual maze, mastering physical tasks. In the real world, sleep deprivation is linked to poor grades in school and low scores on standardized tests. It's difficult to isolate individual components of the sleep cycle, but dreams may be the most important time for consolidating important, long-term memories. In early, non-REM sleep phases, we revisit the immediate past, but later REM stages — when dreams occur — do the most important work, dealing with older, more significant memories. "In non-REM sleep, memory reactivation is directly related to recent experience," Wilson explained. "We find brief snippets of reactivated sequence. I describe it as 'the MTV model of memory': short, edited scenes." In REM, however, "It's not strictly retrieval. It's reevaluation of everything you've learned, including recent and past experience."

The real culprit in the physical and mental consequences of sleep loss may be the deficit of dreams specifically. When people miss out on REM one night, they have extra-long REM periods, or "REM rebound," the next; REM is so crucial that the body rejiggers its natural rhythms to compensate for a shortage. Animals deprived of only REM sleep suffer many of the same symptoms as animals that aren't allowed to sleep at all (although the pace of their deterioration is slower; rats die after about four to six weeks without REM but after about two to three if denied sleep altogether). In one series of studies, William Dement discovered that cats could be kept from dreaming if investigators held them in their laps and gently prodded their noses whenever they entered a REM phase. As the cats racked up days without REM, they grew reckless, manic, even though they were still sleeping. Cats that had ignored rats in their vicinity "ferociously" attacked them.

They "gobbled their food with a rapacity never before observed." They became hypersexual; felines that used to keep to themselves "would continuously try to mount the other cats."

The new sleep science has even made its mark on policy. The campaign for later high-school start times — once just an adolescent fantasy — has been taken up by parents and activists armed with data on the importance of sleep and support from the American Academy of Pediatrics. Sleep deprivation has even been posited as a root cause of poverty. A few years ago, development economists from the University of Pennsylvania set up a "poverty lab" in the Indian city of Chennai and began giving out eye masks and earplugs to people who had to contend with blaring horns and rampant mosquitoes when they lay down at night. Economist Heather Schofield plans to test the effects of these sleep aids on workers' productivity.

As sleep science has blossomed from a niche specialty into a well-funded industry, dream scientists have found homes in growing sleep clinics. By the early 2000s, the stage was set for a renaissance in dream science and a reconsideration of older, forgotten research. And learning about those discoveries can help us understand ourselves more deeply.

chapter 5

PROBLEM-SOLVING

ONE MUGGY SATURDAY IN AUGUST, I WENT ON A date with a man who seemed entirely fine. We drank two beers and went for a walk, and he explained why he liked certain buildings that we passed. We rifled through a box of discarded books on someone's stoop, and he told me why I ought to read William Finnegan's memoir about surfing. We kissed, and his breath tasted like cigarettes; I fought the urge to excuse myself and brush my teeth. We parted ways, and I couldn't muster the energy to answer his emoji-laden follow-up texts about my weekend activities.

The date was mediocre at best—but in the days that followed, I second-guessed my decision not to see him again. Maybe I had written him off too soon; maybe I should have given things a chance to develop. After all, he had some good qualities. He was handsome, tall, employed—and not, refreshingly, as a writer.

It was only after a painfully on-the-nose dream a few weeks later that I stopped doubting my intuition. In the dream, I had agreed to a second date, and I had brought along two friends to observe our interactions and help me assess him. At the end of the group outing, my friends pulled me away and offered a unanimous decision: He was boring. I had made the right call. When I woke up and remembered the dream, I realized two things: I had never liked the guy and, more important, I needed to trust my gut when it came to romantic decisions rather than looking to others to validate what I actually felt.

DREAMS — WHEN WE ARE LUCKY — can give us insight into personal problems and ideas for creative projects. Seeing solutions in surreal metaphors rather than bluntly reasoning them out can help us understand them in a new light. We mull over their symbolism; we marvel at their oddness and reflect on what feels true.

In the 1990s, Deirdre Barrett designed a study to explore how people could use dreams to work through real issues in their lives. She began by teaching a group of college students about the possible link between dreaming and problem-solving and regaling them with stories of dream-inspired discoveries. Next, she asked them to choose a personal issue to tackle in their dreams. Like many undergraduates, they had problems that tended to revolve around relationships or career paths. Every night for a week, they would meditate on the problem for fifteen minutes before bed and then jot down all of their dreams in the morning. At the end of the week, they turned their journals over to two independent readers, who determined that about half of the students had dreamed about their target problem at some point during the week, and a quarter had actually dreamed up a reasonable solution. (The students themselves were even more likely to believe that their dreams contained advice, sometimes in the guise of metaphors that only they could understand.)

An aspiring psychologist was struggling to pick a specialization and had applied to graduate programs in both clinical and industrial psychology. She dreamed that she was on board a plane that was soaring over a map of the United States when the pilot announced that the engine was failing: they needed to find a safe place to land. The student suggested they touch down in Massachusetts, where she had grown up and gone to college and where her family still lived, but the pilot countered that the entire state was "very dangerous," and that they would have to hang on until they reached the West Coast. Upon waking, she recognized something for the first time: Both of the clinical programs she had applied to were in Massachusetts, while the industrial ones were farther away. "I realize that there is a lot wrong with staying at home and that, funny as it sounds, getting away is probably more important than which kind of program I go to." Another student — a woman whose periods had become irregular — dreamed that her doctor told her the problem was her intense exercise regimen and strict diet, which she had neglected to mention to her actual physician. "I guess I should tell him about the diet and exercise, huh?" she admitted when she reflected on the dream.

Barrett's results built on a line of research that stretches back to the nineteenth century, when biologist Charles Child polled his students on whether they included dreams in their decision-making process. In the 1970s, William Dement took a more rigorous tack, giving five hundred undergraduates copies of brainteasers and instructions not to peek at them until just before bed. On each of three nights, they were to open one problem and spend fifteen minutes trying to solve it. In the morning, they wrote down whatever dreams they could remember and attempted the riddle again.

In the first puzzle, students were told that the letters *O, T, T, F, F* marked the beginning of an infinite pattern, and their job was to predict how it would continue. One student dreamed about strolling through an art gallery counting the pictures on the wall: "One, two, three, four, five." When he came to the places where the sixth and seventh paintings should have been, he found only vacant frames. "I stared at the empty frames with a peculiar feeling that some mystery was about to be solved," he wrote the next morning. "I realized that the sixth and seventh spaces were the solution to the problem!" (The correct answer, as Dement explained in *Some Must Watch While Some*

Must Sleep: "The next two letters in the sequence are S, S. The letters represent the first letters used in spelling out the numerical sequence, "One, Two, Three, Four, Five, Six, Seven, etc.")

Dement collected a total of 1,148 dream reports, 87 of which referred to the problem and 7 of which contained the answer. He recognized the limits of this design—the whole thing relied on self-reporting, and the students didn't have a strong incentive to succeed; trying to impress a researcher didn't compare with trying to figure out a career or relationship. Even so, he was convinced "that the dream solutions obtained in this experiment were valid examples of problem solving." In fact, he speculated, "We cannot eliminate the possibility that all of us are presented solutions to our problems quite regularly in our dreams. Perhaps only the most perceptive dreamers possess the ability to recognize a solution that is presented in a disguised or symbolic fashion."

In the 1980s, Morton Schatzman, a psychiatrist in London, enlisted the readers of various British magazines in his own exploration of the link between dreams and creativity. He published a series of puzzles in papers like the *Sunday Times* and the *New Scientist* and challenged readers to dream up solutions and send them in. Again, this wasn't a controlled experiment, but Schatzman's project had the advantage of drawing from a huge pool of respondents, and it yielded a number of intriguing stories. One *New Scientist* subscriber decided to put his dreams to work on this riddle: "What is remarkable about the following sentence? 'I am not very happy acting pleased whenever prominent scientists overmagnify intellectual enlightenment.'" The night after reading the problem, he dreamed that he was delivering a lecture on hypnosis to a group of scientists who, to his annoyance, weren't paying attention. In the morning, he remembered that the audience in his dream had been sitting in a strange configuration: one guest was sitting alone at his own table, two scientists sat at a nearby table, three at the next, and so on. "I began to feel that numbers were important in this problem, and I counted the number of words in the sentence," he wrote. "As I did so, I realized that it was the number of letters in each word that was important," and he had the answer: the first word of the sentence consisted of a single character, the second of two, and so on.

Another woman, a reader of the *Sunday Times,* resolved to dream about this Schatzman riddle: "Which of the following verbs does not belong in this group: *bring, catch, draw, fight, seek, teach* and *think*?" The woman dreamed of watching the British actor Michael Caine pointing backward over his shoulder, and understood that he was performing a pantomime; his gesture was supposed to indicate the past tense. When she woke up and pondered the problem, she realized the answer. "I saw that the only one of the verbs whose past tense doesn't end in *-ght* was *draw.*"

"These examples," Schatzman modestly concluded, "indicate that at least some dreams are not mere mental doodlings, but have meaning and purpose."

When we dream — as when we brainstorm or free-associate — we withhold judgment, allowing ourselves to consider ideas we might otherwise dismiss and to confront emotional truths we would rather resist. In fact, psychologists have even identified qualitative parallels between dreaming and daytime free-associating. Both dreams and daydreams are emotionally and visually intense but only occasionally incorporate senses like taste, smell, and physical pain and pleasure. Both reflect present concerns and anxieties about the future and plunge us into circumstances that are impossible or bizarre. The dreamer, like the daydreamer, lacks "meta-awareness"; she is ignorant of the state she's in and succumbs to the illusion that the fictive world is the only one. Tennyson put it well in "The Higher Pantheism": "Dreams are true while they last, and do we not live in dreams?"

As sleep comes on and mental activity winds down, what neuroscientists call the default-mode network kicks in. "This is a network of brain regions that becomes active whenever you're not actively engaged in some task," Harvard researcher Robert Stickgold explained. This paradoxical flaring-up of the brain was discovered by accident.

In the 1990s, when scientists began using PET scans to learn about human cognition, they planned to treat the resting brain as a baseline, a passive control condition, to be contrasted with the active brain working on whatever had been assigned. In a typical study, subjects would be asked to perform some cognitive task, like reading a short passage or determining the direction of a moving dot, and then lie in the PET machines afterward, resting; the researchers would compare the brain scans from the two parts of the experiment. They assumed that people's brains would shut down as soon as they stopped working on the task — and they were stunned to see that the medial prefrontal and lateral parietal cortices actually showed more activation. "It turns out that when you're doing nothing, your brain is always working," Stickgold said. "You're driving along, you're walking down the street, you're waiting for your waitress to bring you your food." You aren't focusing on anything in particular, but your brain is turning over lingering ambiguities, ruminating on unfinished business.

The default-mode network has since been implicated in modes of thought like mind-wandering, creative thinking, and dreaming. "As you're falling asleep, your brain is falling into that default mode where it's reviewing events from the day," Stickgold explained. "It's reviewing everything that has a tag on it that says, 'You're not done with this.'" That could be anything new, vague or intense — a game of Tetris or a hike up a steep mountain, a confusing conversation or nerve-racking project.

Dreaming can be thought of as a kind of extreme fantasizing. When a team of Harvard neurophysiologists compared descriptions of their students' dreams and waking fantasies, they found that both were bizarre but that dreams had about twice as many bizarre elements, like inexplicable appearances of new characters or abrupt shifts in the story line. On a neural level, dreaming and mind-wandering rely on many of the same mechanisms. In 2013, a team of psychologists led by Kieran Fox compared brain images captured during episodes of spontaneous mind-wandering and during dreams and found a significant overlap; both types of activity involved cognitive regions of the brain like the medial prefrontal cortex and the medial temporal lobe. REM sleep also engaged cortical areas involved in visual processing, leading Fox to conclude that dreaming "can be considered an intensified version of spontaneous waking thoughts, which are only moderately visual in nature."

One of the most important functions of dreaming is to facilitate outside-the-box thinking. Dreams bombard us with plenty of unintelligible scraps, but the odd gem is buried among the junk. "I sometimes say that when we're dreaming, we become venture capitalists," Stickgold said. "We're not interested in safe investments that will give us a five percent return. We're looking for risky investments. It's fine if most of the time it's garbage, because you dream all night long. If eighty percent of it is wasted, and an hour of it is important, otherwise inaccessible associations, that's really valuable."

In a landmark study in 1999, Stickgold discovered that people made looser, less obvious word associations when they had just been woken from dream sleep than when they were fully alert. He had college students spend three nights in the sleep lab and woke them twice each night. At both awakenings, plus once before bed and again in the morning, they would see a word flash on a screen, followed by a series of letters; their task was to figure out, as fast as possible, whether the second string of characters represented a real word or only random gibberish. In some sets, the second word bore a strong relationship to the first; if the first term was *long*, the second might be the direct opposite, *short*. In others, the word pairs were only loosely related, forcing the students to draw more oblique connections: *thief* and *wrong; cowboy* and *rough*. Normally, people were quicker to recognize the second set of letters as a word if it was obviously related to the first. But when they were woken out of REM, that pattern was reversed: the students were better at identifying the weakly linked word pairs. Those distant connections are "a precursor to creativity," Stickgold said, since creativity entails taking "two pieces of information you already had and seeing a novel way of putting them together."

"When you're dreaming — as opposed to ruminating during the day or lying in bed — you are working in a much wider network of associations," Stickgold added. "In REM sleep specifically, you're more likely to activate a distant association than a tight one." The cognitive state of dream sleep is a perfect breeding ground for trying out new connections. Our frontal lobes — the brain's logic centers — go dark, and at the same time, we lose access to the hippocampus, where new memories are stored. Instead of just replaying recent experiences, the dreaming brain reaches back into the memory storage system, where it's apt to land on far-flung files.

If we want to use dreams to find new answers to our problems, we have to remember them. The functions highlighted by Stickgold and Wilson — the role of dreams in learning and memory formation — don't depend on the ability to recall dreams; as interesting as it is to peek into the inner workings of our brains, it isn't necessary. But we can't make the best use of dreams if we forget them.

Dreams, by their nature, are difficult to hold on to. They often lack any kind of cohesion or narrative structure, and a chaotic series of images will always be harder to reconstruct than a tidy story (just as it's harder to remember a string of random letters than a word). Memories tend to be encoded through repetition, but each dream is unique. Psychologist Ernest Schachtel compared the challenges of recalling dreams to the difficulty of remembering childhood memories. Both, he wrote in his 1959 book *Metamorphosis,* involve "experience and thought transcending the conventional schemata of the culture."

Some of the obstacles to dream recall are beyond our control. Men tend to have a harder time remembering their dreams, as do older people — in fact, dream recall usually peaks in young adulthood.

People who frequently remember their dreams may share certain qualities: they tend to score high on psychological measures for "openness to experience" and "tolerance of ambiguity." Some of these traits appear to be ingrained by adulthood, but others can be improved. In one recent study, psychologists noticed that a program meant to stave off dementia had a side effect of making seniors more "open to experience": more curious and creative, more willing to consider new ideas. They didn't expect this personality trait to be malleable so late in life, but by the end of a four-month experiment in which a group of elderly subjects practiced increasingly difficult sudoku and crossword puzzles, those who had participated in the program demonstrated not only gains in the target skills, like problem-solving and pattern recognition, but also an increase in openness to experience. "There are certain models that say, functionally, personality doesn't change after age 20 or age 30," one of the professors who worked on the study said. "But here you have a study that has successfully changed personality traits in a set of individuals who are (on average) 75."

Fortunately, though, dream recall is a skill that most people can improve through minimal effort — no personality change required. For many, the mere desire to remember dreams is enough; reminding yourself of your intention as you fall asleep can yield a bounty of memories in the morning. "The single most important step in encouraging and enhancing dream recall is deciding cleanly and wholeheartedly that you really *are* interested and really *do* want to remember your dreams," Jeremy Taylor, who spent decades leading dream groups, wrote in *The Wisdom of Your Dreams*. "Focusing conscious attention on the desire and decision to recall dreams, particularly as you are falling asleep, will almost always increase the number and quality of dream memories upon awakening." Lifestyle factors can affect dream recall too; it helps not to drink too much before bed, since alcohol suppresses REM sleep. Many people online tout various vitamins and supplements as dream enhancers, but the science is shaky; the most popular dream booster is vitamin B_6, but the hype appears to be based on a single study of twelve college students in which the researchers themselves stated plainly that the experiment was only preliminary.

The easiest, most effective way to boost dream recall is to keep a dream journal and write in it first thing in the morning. When her patients express an interest in working with their dreams, psychologist Meg Jay advises them to start keeping a dream journal. "Any time they wake up, if they write it down, the brain gets better at it. If you do that regularly, you'll go from 'Hey, I don't dream' to 'I remember all three or four of my dreams in a night.'"

In the 1970s, psychologist Henry Reed (who has also identified professionally as a shirt maker and a goat rancher) assigned a group of seventeen college students to log their dreams in a daily journal and attend twice-weekly classes on the meaning of dreams. Over the course of the twelve-week project, the students' memories of their dreams became much more vivid: 58 percent of their entries in the first half of the project included visual details, but in the second half, that number rose to 73 percent. Meanwhile, the proportion of dreams that mentioned colors rose from 33 to 52 percent. And the students enjoyed the improvement so much that they kept up with their diaries even after their teacher had stopped paying attention: three months later, twelve of the seventeen were still recording their dreams.

It's best to write in the dream journal as soon as you become conscious — before making coffee, before looking at your phone, before getting out of bed; even, if possible, before opening your eyes. Any bodily motion or engagement with the physical environment can jolt you out of your internal world and erase your memories from the night. In 2009, a pair of psychologists showed the ramifications of even a brief distraction on dream recall. Amy Parke and Caroline Horton woke their twenty-eight subjects with a phone call and had half of them complete a simple cognitive exercise — like circling all the *e*'s in a passage — before filling out their dream diaries. As Parke and Horton expected, the group that had to perform the task before transcribing their dreams gave shorter, less detailed dream reports.

It doesn't matter whether you write your dreams in a notebook, type them up on your laptop or phone, speak them into a voice recorder, or draw them on a sketchpad. Consistency is more important than method; whichever technique is easiest to stick with is best. I used to keep a pen-and-paper dream journal, marking down the following day's date the night before; setting up the page would reinforce my goal and make it even easier to write down my dreams first thing in the morning. More recently, as I've been remembering more details from my dreams, I've been keeping a log on a Word document. There are downsides to opening my computer first thing in the morning, but it's more realistic to type up hundreds of words than to write them out by hand, and it's easier to look for patterns on a typed page. Of course, handwritten entries can be typed up, and voice memos can be transcribed. The dream journal is a document that should be reread and referred back to, engaged with as much as possible, searched for repeating themes and reflections of real-life experience.

If these methods aren't working, Jeremy Taylor has devised another strategy that can help. If you wake up with no recollection of your dreams, you might resuscitate a memory by reenacting the poses you held in your sleep. "We all have a series of body postures that we habitually roll through over the course of a night's sleep," he wrote. "Returning to each of these postures in turn is likely to release dream memories — presumably memories of dreams that occurred while the dreamer was lying in those positions." And if that fails, you might try to jog your memory by visualizing the faces of people whom you have strong feelings about, since these are the characters most likely to populate your dreams.

Waking up naturally can help you hold on to your dreams; psychologist Rubin Naiman compared waking up with an alarm to "being abruptly ushered out of a movie theater whenever a film was nearing its conclusion." If you do have to set an alarm, it's best to time it to go off at the end of a REM stage (so a multiple of ninety minutes after going to sleep). When researchers wake people during REM sleep, they can usually remember their dreams; the more time that elapses between a REM period and waking up, the less likely you are to remember your dream. (Dreaming is most frequent during, but not entirely limited to, REM phases; according to neuroscientist Mark Solms, "The strongest claim is that 90–95% of awakenings from REM sleep produce dream reports, whereas only 5–10% of awakenings from NREM sleep produce equivalent reports.") Thanks to the same logic, waking up at strategic intervals throughout the night — toward the end of REM phases — can maximize the number of dreams you recall; if you sleep for eight hours, you might set an alarm to go off after three REM cycles (about four and a half hours) or four (about six hours into the night). Robert Stickgold recommends a more natural method: drink a couple of glasses of water right before bed.

Every time you wake up — even in the middle of the night — you should record notes on the most important elements of your recent dream; if you fall back to sleep without recording it, the memory is likely to be gone by the time you wake from the next dream. Even brief bullet points from the middle of the night can trigger detailed memories of the dream the next day.

These methods may mean a tradeoff with sleep quality, at least until the habits become ingrained. "Most of the people who don't remember their dreams are people who fall asleep quickly, sleep soundly through the night, wake up with an alarm clock and jump quickly out of bed," said Stickgold. "They don't have those times where they're just sort of waking up."

THE CREATIVE POWER of dreams has been exploited to especially dramatic effect by artists and inventors. The list of artistic feats attributed to dreams spans practically every arena of human accomplishment, from literature, visual arts, and music to science, sports, and technology. Beethoven and Paul McCartney cited dreams as the spark behind some of their musical compositions (including McCartney's famous "Yesterday"). Some of the most recognizable sequences in film — sections of Ingmar Bergman's *Wild Strawberries,* Fellini's *8½,* Richard Linklater's *Waking Life* — are translations of the directors' dreams. Mary Shelley credited dreams with inspiring *Frankenstein;* E. B. White with *Stuart Little.* Some scholars believe that the oldest art in the world was inspired by dreams. "What if the artists who painted the magnificent caves of Lascaux and Corvet were history's first dream recallers, and the paintings on the cave walls were their dream journals?" asked Kelly Bulkeley. Many of the scenes shown in cave paintings — hordes of humanoid beasts, wild animals layered on top of pointillist patterns — have a surreal, dream-like quality. And humans, like the Greeks seeking diagnostic dreams, have often gone into caves to incubate important visions.

Both dreaming and imaginative thinking involve a kind of abandon, a letting-go. The artist, like the dreamer, gives in to the immediacy of her vision, forgoing her physical environment in favor of the fantasies in her mind's eye. In dreams, as in creation or free association, we indulge irrational thinking and briefly transcend the logic we follow in the day. In her book *Philosophy, Dreaming and the Literary Imagination,* literary scholar Michaela Schrage-Früh went as far as suggesting that humans began writing fiction as a way of sharing and making sense of their dreams. "The earliest documents of written-down stories actually record and interpret the author's dreams, for example, the Sumerian dream text *The Epic of Gilgamesh,* inscribed on a clay tablet over 5000 years ago," she pointed out. "The need to tell stories may have been sparked not least of all by the desire to communicate one's dreams."

Exceptionally creative people may be naturally prone to vivid dreaming; high dream recall is correlated with habits and personality traits often shared by artists, such as "openness to experience," "tolerance of ambiguity," an inclination toward fantasy, and a tendency to daydream. People who remember their dreams every night tend to have an easier time of losing themselves in projects and are likelier to agree with statements like "I am full of ideas" and "I am interested in abstractions."

For several years in the 1990s, physician James Pagel interviewed screenwriters, actors, and directors at the Sundance film festival about how dreams figured in their daily lives—how often they looked to dreams for artistic inspiration or insight into personal problems. "They blew all the scales," said Pagel, who now runs a sleep-disorder clinic in Colorado. These professional creatives' dream recall was almost twice as high as the general population, and they used their dreams in their work "all the time. It was very rare to find someone who didn't use their dreams." Different types of artists had figured out how best to harness the generative potential of their dreams for their own genre. "Screenwriters often used incubation, where you go to sleep visualizing a problem in your creative process," he said. "They would go to bed with that in their head and wake up in the morning with a new idea for their screenplay. Actors blew the scales across the board. They tended to use their dreams not just in their creative process, but in all aspects of their life—in relationships, in making decisions, in relating to self and others on all different levels."

Wondering about the other end of the spectrum, Pagel next went looking for people who never remembered their dreams. He wanted to find out if they had anything in common—a cognitive deficit, perhaps, or some unusual character trait or habit. A surprisingly high number of people who sought treatment at his clinic—between 6 and 9 percent—claimed that they never dreamed, but most would surprise themselves by reporting dreams if they were woken up at the right time, or they would recall one from childhood if Pagel probed. It took him years to scrounge up enough people for a study, but Pagel eventually managed to find sixteen people who really did appear not to remember their dreams at all, about 1 out of every 262 patients. There was nothing outwardly wrong with these people. They had families and held down jobs — one was a math professor — and exhibited no obvious psychological symptoms.

They shared only one idiosyncrasy. Almost everyone who passed through his lab had some kind of creative outlet in life — crafts, sports, music — except the non-dream-recallers. "I think one of the major reasons we have dreams is for their use in creativity," Pagel said. "Creativity is one of the primary survival characteristics of our species." As long as most people remember their dreams, the odd non-dream-recaller won't doom the human race; an individual "can function well in our society without a creative process, but as a species, we require the capacity to develop alternate approaches to problems."

Artists may have a built-in head start, but many hone their dream recall by keeping dream journals, practicing dream incubation, or even teaching themselves to lucid dream. In the *New York Review of Books,* poet Charles Simic described a writer friend who made a habit of eating an entire pizza at midnight, setting an alarm, and dragging himself out of bed at four a.m. to transcribe his dreams. When the sheltered Charlotte Brontë wanted to write about something she had never experienced — like smoking opium — she willed herself to dream about it; Salvador Dalí and Robert Louis Stevenson were skilled lucid dreamers. Dalí devised his own technique, which he shared in *Fifty Secrets of Magic Craftsmanship,* his guidebook for budding painters: Sit down for a nap with a heavy key in hand. When you nod off, the key will drop. That sound will wake you up, and you will have access to the images — the "hypnagogic hallucinations" — that were whirling through your mind as you drifted off. Whether story-like or absurd, whether obviously relevant or not, dreams remind artists that, even when they feel blocked, the ability to create fictional worlds still resides within them. In dream journals, writers can scribble and free-associate unconstrained by the specter of publication. The scratching of the pen on paper or the comforting clacking of the keyboard in the morning — the trappings of industriousness, of words effortlessly spilling forth—can ease the transition into the kind of writing that takes more work.

BRITISH NOVELIST GRAHAM GREENE started keeping a dream journal when he was sixteen. He had been the target of bullies at his posh boarding school, and his childhood uneasiness morphed into full-blown depression. As a teenager, he poured his nascent creative talents into devising interesting, yet ultimately ineffective, ways to kill himself. He sliced a vein in his leg; ingested alarming quantities of conventional toxins, like allergy pills, and more romantic ones, like deadly nightshade; went swimming alone with a stomach full of aspirin. His family finally threw up their hands and sent him off to London for a course of psychotherapy—"an astonishing thing in 1920," he later wrote. At the urging of his analyst, Greene agreed to write down his dreams every morning. He graduated from therapy and returned to school after a few months, but he maintained the habit, on and off, for the rest of his life, keeping a pen and pad beside his bed and jotting down his dreams every time he woke up, sometimes as often as four or five times a night.

Yvonne Cloetta, Greene's mistress, later recalled how he would structure his entire workday to harness the creative power of his dreams. He would draft new sections of a novel in the morning—tapping away until he had met his self-imposed daily quota of five hundred words—and then read them over before bed, "leaving his subconscious to work during the night."

"Some dreams enabled him to overcome a 'blockage'; others provided him on occasion with material for short stories or even an idea for a new novel," she wrote. "'If one can remember an entire dream,'" she remembered Greene saying, "'the result is a sense of entertainment sufficiently marked to give one the illusion of being catapulted into a different world. One finds oneself remote from one's conscious preoccupations.'"

As Greene neared the end of his life, he authorized excerpts of his dream diaries for posthumous publication; once he had escaped the possibility of mortal embarrassment, he would allow his fans a peek into the inner life that had inspired his two dozen novels. Many of the entries in *A World of My Own: A Dream Diary* read like fully realized short stories. In one, he found his mother's corpse in bed; when he went to lift her body, she opened her mouth and complained that she was cold. In others, he traveled to Sydney and Sierra Leone and communed with his literary heroes, fielding criticism from a mustachioed T. S. Eliot and compliments from a friendly D. H. Lawrence.

Greene's fixation on dreams is palpable in his work. In *The End of the Affair,* dreams are a source of both pain and inspiration for Maurice Bendrix, a novelist struggling to make sense of an old romance with a friend's late wife. Looking back, Bendrix resents the dreams he endured after his lover abandoned him. "I remember I dreamed a lot of Sarah in those obscure days or weeks. Sometimes I would wake with a sense of pain, sometimes with pleasure. If a woman is in one's thoughts all day, one should not have to dream about her at night." But in spite of the agony dreams can inflict, Bendrix still relies on them; in the very same paragraph, he extols the importance of sleep in the mysterious creative process. The writer goes about his day, "preoccupied with shopping and income tax returns and chance conversations, but the stream of the unconscious continues to flow undisturbed, solving problems, planning ahead," until finally "the words come as though from the air . . . the work has been done while one slept or shopped or talked with friends."

A dream can serve as a signpost that a project is going well or a warning that it's time to give up. Whenever Maya Angelou dreamed of discovering a skyscraper under construction and scaling it, scaffolds and all, she knew that her writing was on the right track, that she was "telling the truth and telling it well." "I have no sense of dizziness or discomfort or vertigo," she told Naomi Epel, who interviewed authors for her book *Writers Dreaming.* "I'm just climbing. I can't tell you how delicious it is!"

Writer Kathryn Davis, however, credited a dream with helping her concede that she should give up on her first attempt at a novel. Flush from the success of her short stories, she had decided to try her hand at a longer piece of fiction. She struggled with the project — a historical novel about Eskimos at the World's Fair — but she knew that all writers had their doubts, so she did her best to push hers aside. "When you're working on a novel, you have this idea that it's not easy to write one, and that one of the things you have to do is persist through the difficult times," she said later. A vivid dream forced her to rethink that assumption. "In the dream, I was walking out of the house toward the barn, and this horse stuck its head out through the top of the barn door," she said. "It was Mr. Ed, the talking horse. He said to me, 'It's borrrring,' and I knew he meant my novel. From that moment on, I put my novel in a box and moved on to the project that became my actual first novel *Labrador*" — which garnered rave reviews and launched her career. "That was all because of Mr. Ed."

STEPHEN KING HAD written several hundred pages of the novel *It* when he lost his momentum. He was already a best-selling writer, but this — the terrifying tale of a child-stalking clown — was his most ambitious project yet. "I had a lot of time and a lot of my sense of craft invested in the idea of being able to finish this huge, long book," he told Epel. Now he was edging closer and closer to a plot point he could not see beyond, and the prospect filled him with dread. He feared that the entire project could be derailed. "I didn't know what was going to happen," he said. "And that made me extremely nervous. Because that's the way books don't get done." With that ominous thought hanging over him, he fell into bed one night, berating himself: "I've got to have an idea. I've got to have an idea!"

A few hours later, he was roaming through a garbage dump piled high with old refrigerators, and he knew instinctively that he had been deposited into the scene that was giving him so much trouble. He set out to explore the strange landscape; he approached one of the scrapped refrigerators and pried open the door. A swarm of macaroni-shaped objects dangled from corroded shelves, and one of them flew out and perched on his hand. His arm was instantly flooded with a feeling of warmth — "like when you get a subcutaneous shot of Novocain" — and he realized that the thing was sucking his blood. "Then they all started to fly out of this refrigerator and to land on me," he recalled. "They were all these leeches that looked like macaroni shells. And they were swelling up." When he jerked himself awake from his nightmare, he was, naturally, "very frightened" — but he was also "very happy": he knew he had the plot device he'd been searching for. He wrote up the dream and transferred it directly into the book, unaltered. And, of course, he finished *It* — which clocked in at over one thousand pages, fulfilling all of his epic ambitions, terrorizing millions of readers, and putting clowns across America out of work. "I really think what happened with this dream," he reflected, "was that I went to sleep and the subconscious went right on working and finally sent up this dream the way that you would send somebody an interoffice message in a pneumatic tube." King was no stranger to the inspirational capacity of dreams or the parallels between dreaming and writing. "I've always used dreams the way you'd use mirrors to look at something you couldn't see head-on — the way that you use a mirror to look at your hair in the back," he said. "Part of my function as a writer is to dream awake" — to trick himself into a trancelike state where stories and images come and go unbidden.

EVEN DOWN-TO-EARTH SCIENTISTS and mathematicians can benefit from the creative jolt of a dream. In 1902, a young German physiologist named Otto Loewi caught wind of the debate over how nerves communicated with muscles — did they transmit signals electrically, or did they release chemicals that traveled through the body? When Loewi learned about the parallels in how nerves and certain drugs affect muscles, he decided he favored the less popular theory of chemical transmission — but he couldn't imagine how to test his intuition, so he pushed it to the back of his mind and carried on with his research on animal metabolism.

Nearly twenty years later, a solution to that forgotten problem came to him in the middle of the night, apparently out of the blue. "In the night of Easter Saturday, 1921, I awoke, turned on the light, and jotted down a few notes on a tiny slip of paper," Loewi later wrote. "Then I fell asleep again. It occurred to me at six o'clock in the morning that during the night I had written down something most important, but I was unable to decipher the scrawl. That Sunday was the most desperate day in my whole scientific life." The following night, though, he got lucky: the dream came back. "I awoke again, at three o'clock, and I remembered what it was," he wrote. "This time I did not take any risk: I got up immediately, went to the laboratory, [and] made the experiment."

As if in a daze, Loewi began acting out the scene from his dream. He dissected two frogs, extracted their hearts, and submerged them in saline — in which the organs would continue to beat outside the body. Next, he found a battery and used it to stimulate one heart's vagus nerve, which — as he expected — slowed the rate of its beating. He took a portion of the saline solution from the decelerated heart's container and transferred it to the other jar. The second heart also slowed down, confirming Loewi's old hunch: some chemical generated by the vagus nerve, rather than the electric charge from the battery, was responsible for the heart's change in pace. "At five o'clock the chemical transmission of nervous impulse was conclusively proved," he wrote. He had identified the first neurotransmitter and laid the groundwork for the entire field of neuroscience.

Loewi's dream — and its rich real-life fruits — persuaded the rational researcher that "we should sometimes trust a sudden intuition without too much skepticism."

"Consciously I never before had dealt with the problem of the transmission of the nervous impulse," he wrote. "It therefore will always remain a mystery to me that I was predestined and enabled to find the mode of solving this problem, considered for decades to be one of the most urgent ones in physiology." If he had come up with the idea in the harsh light of day, he suspected, he would have picked it apart and dismissed it.

Academics like Loewi tend to guard their theories closely. In the competitive world of academia, citations are currency to be converted into scarce jobs and promotions. Yet when it came time to put together the acknowledgments for a paper he wrote in the 1960s, MIT mathematician Donald Newman was oddly generous, giving credit to a figment of his own dream. Newman and his friend John Nash often talked about numbers and proofs, chatting about whatever they were working on; mathematics was at the core of their long, intense friendship. But the key to this particular problem—the conversation that had led to the eureka moment—took place inside Newman's own head. Stuck on a question, Newman dreamed that he and Nash were dining at a restaurant in Cambridge. Newman asked Nash what to do. When Newman woke up, he had the answer. "It was not my solution," Newman told PBS in 2002. "I could not have done it myself."

"I could not have done it myself": Newman's humble claim captures what's so alluring about the prospect of finding inspiration within a dream. It's a romantic idea; it fits with our conception of creativity as an inexplicable force, a mysterious type of work. It suggests the seductive possibility of short-circuiting the arduous creative process, of dreams as a meeting place for the muses.

chapter 6

PREPARATION FOR LIFE

NOT LONG AGO, I HAD A SERIES OF UNCANNY dreams that explicitly referred to an upcoming event that I was nervous about. I was raised as a vegetarian, and for the first two and a half decades of my life, I did not — to the best of my knowledge — consume a single morsel of meat or fish or even a vegetable that had been in contact with a dead animal. I held my breath when I walked past meaty restaurants. In high school, I made a boyfriend brush his teeth after eating meat in front of me.

But after spending nearly a quarter-century carefully avoiding meat, I decided I'd had enough. I never had any real ideological justification for my diet; I've never felt any particular affinity with animals, and I consider my health to be good enough. My diet was based on nothing more than a very ingrained habit. I was tired of dealing with an annoying restriction when I traveled to places less veggie-friendly than Brooklyn, and I was tired of inconveniencing friends and hosts. At one dinner party, a friend cooked a separate vegetarian meal for me — then absent-mindedly added fish sauce and had to prepare a third dish. I wondered why I was putting her to all this trouble. Plus, I was curious what I might be missing. Meat seemed to be popular.

My friends planned a Meat Party to welcome me to the omnivorous fold. Fish tacos, they promised, would provide a gentle entrée into the world of meat; the mild fishy taste would be disguised by stronger flavors, like salsa and broccoli, and I'd wash the whole thing down with copious amounts of alcohol. I gave myself one month to mentally prepare, and in the run-up to the big day, I had three vivid dreams about eating meat.

> I'm at a buffet. The main dish is a sort of flubbery chicken: like sticks of rubber. I decide to eat it; I make it through one or two pieces. I announce to the room that I am eating meat. No one is impressed.

> I try some sort of cracker; it doesn't taste good. Then I realize that it's flavored with bacon.

> I'm eating salami from a big bowl of worm-like things. It tastes slimy but also fine. It is anti-climactic. Suddenly, I wonder if salami is kosher. I Google it, and find that the Reform Union says it's okay.

From each dream, I woke feeling a little more prepared, a little more certain that I was ready. When I finally lifted that first bite of tilapia to my mouth, I was apprehensive and slightly grossed out. But then I swallowed, and the strongest feeling I had was déjà vu.

According to the threat-simulation hypothesis, dreams evolved to serve an important psychological function: they let us work through our anxieties in a low-risk environment, helping us practice for stressful events and cope with grief and trauma. Most of the emotions we experience in dreams, as Finnish scientist Antti Revonsuo noted in the 1990s, are negative; the most common ones are fear, helplessness, anxiety, and guilt. For an evolutionary psychologist like Revonsuo, this presents a puzzle. Why would our minds subject us to something so consistently unpleasant? If our ancestors could practice dealing with dangerous situations as they slept, he reasoned, they might have an advantage when it came time to confront them in the day. Early human life was a minefield of wild animals, unpredictable terrain, and hostile fellow humans; any edge might improve a person's chances of survival. This theory explains the prevalence of negativity and aggression in dreams as well as the primal nature of many dream environments; even city-dwellers with little experience of the wilderness or interpersonal violence often dream about being attacked by dangerous animals or ominous strangers. Activities like reading and writing—relatively recent developments in human history—are more unusual in dreams.

As the threat-simulation hypothesis would predict, animals appear to dream about survival-related activities like hunting, fighting, and eating. If deprived of REM sleep, animals can't perform even the most basic tasks. In 2004, a group of psychiatrists at the University of Wisconsin–Madison designed an experiment to explore how dream deprivation affected rats' ability to respond to threats. They used the flowerpot technique—a deceptively pleasant-sounding term for what is really a form of rat torture—to deprive some of the rats of REM without altering the total amount of time they slept. Each rat was placed atop its own upside-down flowerpot floating in a jar of water. When the rodent got tired, it would lie down on top of the flowerpot and fall asleep. As soon as it entered REM, its muscles would freeze and it would fall off its perch, land in the water, and wake up. The rat would then crawl back up onto its pot and lie down again, but the cycle would repeat, preventing it from ever catching more than a moment of REM sleep.

After several days, these unfortunate rodents, as well as some rats that had been allowed to sleep normally, were put through their paces. They were confronted with strange objects, dropped in an empty tank, deposited into a maze, and exposed to electric shocks. In each scenario, the rats that had been allowed to rest responded appropriately while the REM-deprived group behaved recklessly. Faced with a foreign object, they groomed themselves instead of burying it. Forced to navigate a maze or an empty tank, they gravitated toward the dangerous open areas rather than sticking to the darker fringe (where, in the wild, they would be less likely to encounter predators). They failed to freeze after receiving an electric shock. Even when the REM-deprived rats were given amphetamines, which can temporarily reverse the effects of sleepiness, their behavior didn't improve; it was the loss of dream sleep specifically, not just general tiredness, that made them self-destructive.

Unless we are contestants on *Survivor* or *The Hunger Games,* threats in modern life tend to be less dramatic than a life-or-death maze, and we have the anxiety dreams to match. The exam dream — in which a dreamer finds herself woefully unprepared, and perhaps underdressed, for an important test — is the prototypical modern-human version of the racing-rat or hunting-cat dream. Even if she fails in the dream, the test seems familiar in real life — and the illusion of familiarity can translate into a real advantage. My own exam dreams are simultaneously boring to reproduce and pulse-quickening to remember. "I'm leaving the test center and realize I forgot to write any essays." "I'm taking my exams and then I remember I'm not wearing pants." These nightmares come years after I've left school, but they pop up when I'm worried about something else, like a looming deadline or the reception of a story I'm about to publish. Some psychologists believe that in times of stress, we dream about exams that we actually succeeded in; our brains are reminding us of a time when we prevailed over something we had feared, boosting our confidence in real life. In his own anxiety dreams, Freud relived his botany and chemistry exams, which he had passed, rather than the medical-law exam he had failed. In reality, I wore clothes to my college finals and didn't leave them blank. But mulling over the worst-case scenarios in the light of day forces me to recognize how unlikely, even ridiculous, they are; confronting them saps them of their power to terrify, even makes me laugh. I wake with a sigh of relief — no matter how unprepared I feel for this meeting, at least I won't turn up naked. However poorly this article lands, I won't have to reapply to college. Whatever my editor thinks of my draft, he isn't going to send it to all of my exes for comment. Nor is he likely to invoke a secret demon clause that allows him to impersonate a demon and cancel my contract.

In 2014, researchers from the Sorbonne, led by neurologist Isabelle Arnulf, contacted thousands of aspiring doctors on the day they were scheduled to take their medical school entrance exam. Nearly three-quarters of the students said they had dreamed about the exam at least once over the course of the semester, and almost all of those dreams had been nightmares: they got lost on their way to the test center, found it impossible to decipher the test questions, or realized they were writing in invisible ink. When Arnulf compared students' dreaming patterns to their grades, she discovered a striking relationship: students who dreamed more often about the test performed better in real life. Most suggestive, all of the top five students had encountered some exam-related obstacle in their dreams, like sleeping through their alarm or running out of time. "Negative anticipation may serve to optimize daytime performance, as in how a chess player imagines all possible moves, particularly the moves leading to a loss, before selecting and playing the better move," she wrote. "The contrast between the horrible situations experienced in dreams (appendicitis, lateness, impossibility of competing) and the more casual reality (good health, appropriate timing and tools) the next morning may desensitize the students to anxiety, which can be reassuring and beneficial for competition."

Failing an exam is just one of a number of classic dream scenarios that seem to recur all over the world. For millennia, people have written and wondered about dreams of flying and of falling, of being naked in public and of losing their teeth. The universality of these dream motifs suggests that there is something fundamental about them — that, like language and music and social groups, they serve some deep-seated evolutionary purpose. Freud suggested that dreams of flying were rooted in childhood memories of being rocked; Jung connected them to conquering challenges in real life. German psychiatrist Michael Schredl — whose conclusions are based on empirical studies rather than guesses — has found that dreams of flying "reflect positive emotional states experienced in waking life."

Another vigorous discussion — which has spanned Egyptian papyri, Vedic scriptures, and journal articles on Japanese college students — concerns the common yet mysterious dream of tooth loss. (It's difficult to know exactly how typical this particular nightmare is, but one study found that 21 percent of American undergraduates had dreamed of their teeth falling out; the dreams are often associated with stress in real life.) According to Bar Hedya, an ancient Jewish mystic, dreaming of losing one's teeth represented a warning that a relative would soon die; the Navajo also considered this nightmare a warning to the dreamer's family. Freud, predictably, saw it as a symbol for castration. My favorite psychoanalytic take — proposed by a Hungarian therapist named Sandor Lorand — is that it represents a wish to return to the toothless, sexless state of babyhood. Scientists have conceded that it's useless to try to interpret an isolated element of a dream without context on the dreamer's life and dreaming patterns — individual dream languages are too diverse — but the tradition continues online; well-trafficked dream-dictionary sites tell me that someone who loses his teeth in a dream fears aging, has misspoken, or has "some hardship in life."

In 1984, a pair of psychologists conducted an empirical study on the topic. Hoping to understand why certain people were prone to these dystopian dental dreams, they rounded up fourteen people who had recurrent nightmares of losing their teeth — "some disappeared, some were pulled out, some were knocked out, some crumpled" — and fourteen who had recurrent dreams of flying.

They found minor personality differences between the two groups — the tooth-loss group scored higher on scales of anxiety and depression — but the main finding was anticlimactic: members of the teeth-dreaming group spent more of their waking time thinking about teeth. Still, their hypothesis about the mechanism underlying this common dream is worth considering. "One possible explanation is that the dream about loss of teeth represents an unconscious, historical vestige or archetype of a time when teeth played a significant role in the lives of early people," they suggested. "The actual loss of one's teeth may have portended death due to resulting changes in diet and other difficulties associated with eating or defense. For the present group of dreamers it seems possible that psychological conditions of helplessness or loss of control may trigger the vestigial, archetypal dream of teeth-loss." There's often something primitive about dreams, featuring settings (like the wilderness), activities (like dodging violent strangers), and perhaps body parts (like teeth) that were once crucial for human survival. In modern society, where survival and success depend more on psychological adjustment, those relics appear in different contexts but perhaps in service of a similar evolutionary function.

A more recent wave of research has found a strong link between dreaming about a new or unfamiliar skill and improving at it in real life. A pair of Brazilian neuroscientists borrowed Harvard psychiatrist Robert Stickgold's experimental setup for a new project, replaced Tetris with a violent video game called Doom, and unveiled their results at the 2009 Society for Neuroscience conference. Sidarta Ribeiro and André Pantoja assembled a group of twenty-two volunteers, taking care, like Stickgold, to include both amateurs and experts, and had them practice Doom — a first-person-shooter game that encourages players to slay virtual monsters with pistols and chainsaws — before bed. Their dreams afterward were as gory as the game.

In the morning, they played again. When Ribeiro and Pantoja analyzed their subjects' dream reports alongside their performance, they discovered a clear link: players who dwelled on Doom in their dreams were more likely to skillfully slaughter their virtual enemies and drive their avatars around hurdles in the morning. EEG recordings provided another window into the role of dreams in real-life gains. In both daytime play and dreams, the novices, who struggled at first with the physical maneuvers, showed greater activity in the brain regions associated with moving their hands, whereas the experts had more activation in the frontal areas of the brain, which are responsible for complex reasoning and decision-making. One potential explanation is that the dreams centered on whatever aspect of their technique needed the most improvement, whether that was basic steering or more advanced strategizing. (Another theory, as Ribeiro explained in an interview with *New Scientist,* was that those "who dreamed about Doom most could simply have been those most motivated to improve.")

The survival-themed game of Doom is a more plausible stand-in for real life than a puzzle like Tetris, and cognitive neuroscientist Erin Wamsley has managed to take it even further, using Stickgold's paradigm to study the role of dreams in mastering everyday activities like playing sports and navigating new territory. "If you give humans an engaging learning task, you can see that learning task showing up again in their dreams," Wamsley explained. "The more people dream about what they just learned, the better their memory is."

In one experiment, Wamsley, who runs her own sleep lab at Furman University in South Carolina, taught forty-three undergraduates to play an arcade game called Alpine Racer, which simulates a downhill ski course. Students would spend hours standing in front of a screen on fake skis, dodging virtual obstacles. On the first night, nearly half of their dream reports incorporated some element of the game, whether directly (one participant saw flashes of a particularly difficult corner) or indirectly (another dreamed that she was running a race through the city of San Francisco).

For another study, Wamsley had ninety-nine undergraduates report to her lab around noon and spend forty-five minutes trying to navigate a three-dimensional maze. Half of the students then retreated to a dark room to nap; the others spent the afternoon hanging around the sleep lab and complying with Wamsley's periodic inquiries into "everything" that went through their minds. At 5:30, each participant gave the maze another try. Again, Wamsley found a powerful link between dreaming about the maze and actually getting better: the students who had dreamed about the task improved an extraordinary ten times as much as those who didn't. Some of their dreams explicitly referenced an element of the maze task, while others recalled more tangential experiences; one student dreamed about a long-ago visit to a labyrinthine cave. Among those who stayed awake, though, "maze-related mentation" — conscious thinking about or replaying of the task — had no effect on performance. Something special seemed to be happening while they dreamed, something they couldn't replicate when they were awake.

Dreams can help people prepare for problems more serious than a made-up maze or the exploration of a benign food group. In 1971, a trio of psychologists led by Louis Breger decided to study the dreams of hospital patients slated for surgery. The stress of awaiting an invasive procedure can turn even the most nonchalant into a nervous wreck, as the patient — already vulnerable and weak from illness — anticipates laying his unconscious body on the table. The patients' preoperative dreams dwelled on those anxieties and fears, in symbols and metaphors if not literally. Tellingly, those who claimed not to be worried about the surgery — who exhibited, in the psychologists' lingo, "strong tendencies toward repression" — actually dreamed about the upcoming procedures more often. Al — an outgoing veteran in his sixties — denied feeling nervous about his vascular surgery; he even claimed that he was incapable of fearing pain because he had never experienced any. Yet his nightmares, littered with broken knives and blocked-up sewers, told a different story. In his dreams, his body was under threat; he stood before a crowd of strangers who looked "like they would like to cut" his throat. As the date of his operation drew closer, his dream-self became more assertive; he practiced taking charge. In one dream, he mended a broken stove. In another, he cleaned out a set of clogged pipes, which Breger compared to Al's own blocked blood vessels. "Al's dreams represent the threat of surgery primarily in symbolic or indirect forms," Breger wrote. Defective appliances weren't a typical feature of Al's nighttime landscape; they appeared in only about 15 percent of his postsurgery dreams, but more than half of his preoperative dreams.

If dreams are so important, and if so many of their functions depend on our understanding them, why do they often seem incomprehensible? Why do they traffic in garbled metaphors and disjointed images? Why should Al dream about clogged pipes rather than surgery?

It could be, as Freud might have suggested, that he wasn't ready to think about his operation directly. Or the dream might have done him a favor by making it more interesting to think about: mysteries are always more compelling than straightforward lessons.

We tend to focus on and discuss dreams that are strange, but most dreams are less bizarre than we might assume. After analyzing more than six hundred dream reports from labs in Brooklyn and Bethesda, psychologist Frederick Snyder concluded that "dreaming consciousness" was in fact "a remarkably faithful replica of waking life." In his sample, 38 percent of the settings were real places that dreamers recognized from their lives; another 43 percent resembled places they already knew. Just 5 percent were considered "exotic," and less than 1 percent counted as "fantastic." When Snyder scored each dream report on various measures of coherence — "Does the narrative hold together as a story?" "Are the events conceivable, even if unlikely?" — he found that, even among the longest dreams, half lacked so much as a single bizarre element, and as many as nine out of ten "would have been considered credible descriptions of everyday experience."

Nonetheless, the fact remains that on some nights, our brains spin scenes that are far-fetched at the very least. In the 1990s, a trio of researchers — Robert Stickgold, Allan Hobson, and Cynthia Rittenhouse — set out to study whether there were any constraints on the dreaming imagination. After analyzing ninety-seven discontinuities in two hundred of their students' dreams, they found that there were, in fact, rules and patterns at play; dreams aren't governed by the normal laws of physics, but they aren't a free-for-all either. What Stickgold called intra-class metamorphoses were much more common than inter-class ones; that is, when a dream-character transformed, it would usually assume the guise of another character rather than an inanimate object (and vice versa). And even within classes, the transformations were far from random: a pool became a beach; an uncle turned into a neighbor; a car changed into a bike. "The most surprising and novel finding of this study is that many discontinuous dream images are paradoxically coherent," the authors wrote. "A dream object does not transform randomly into another object, but into an object that shares formal associative qualities with the first."

In another study, philosopher Antti Revonsuo and psychologist Christina Salmivalli analyzed hundreds of their students' dreams and discovered that the emotions in dreams were usually appropriate to the situation, even if the situation itself was odd. And another important element was strikingly constant: the dreamer's own self was "well preserved" and "rarely plagued by features incongruous with waking reality." "The representation of self is presumably one of the fundamental cornerstones of our long-term memory systems," they explained. Even in dreams, we know who we are.

TO APPRECIATE THE role of dreams and sleep in emotional health, consider what happens when we miss out on them. One study found that preteens with trouble sleeping were more likely to consider suicide in adolescence. Another study of people over sixty-five showed that over a ten-year period, poor sleep raised the suicide risk by 34 percent. Scientists don't fully understand the relationship between sleep and mental health, but some believe that changes in sleep and dreaming aggravate depression. Tiredness makes us paranoid; a good night's REM helps us accurately interpret social cues, whereas a lack of dream sleep leaves us apt to assume the worst. In 2015, Matthew Walker and his team at the University of California, Berkeley, had eighteen young adults view images of faces conveying different emotions, ranging from friendly (eyes relaxed, the corners of the lips upturned) to foreboding (brow furrowed, lips puckered in a thin line), and interpret their expressions. The volunteers performed this task on two separate occasions, once after a full night's sleep and once after being forced to stay awake for twenty-four hours. When they were allowed to rest normally, they had no trouble understanding the models' facial expressions. But when they had been deprived of sleep, the same participants lost the ability to discern different emotions and tended to judge the models as more hostile than they really were.

One of the most dramatic shifts in the sleep of the depressed is a decrease in dream recall. If awoken at an opportune point in the sleep cycle, a mentally healthy person will report a dream about 80 to 90 percent of the time. In severely depressed populations, that number can drop as low as 50 percent, and the dreams people do recall are less vivid; they're shorter and less emotional and involve fewer characters. Whether it's a cause or an effect, the paucity of dreams may deepen the depression, depriving the person of an opportunity to process her pain.

Several of the contributors to *Unholy Ghost: Writers on Depression* looked back on formless, shallow dreams as markers of their darkest periods. Writer Virginia Heffernan "dreamed of slicing, shape-shifting light." Novelist Lesley Dormen's dreams "were filled with water." As William Styron spiraled into suicidal depression, he slept in restless fits and starts, and his dreams — which had always been elaborate — grew dim. "I think one has to have dreams in order to fulfill the function of sleep," he once said. "And so to have an absence of dreams was an almost intolerable aspect of my illness. I was quite conscious of the fact that when I woke from a kind of synthetic, pill-induced sleep (I was taking tranquilizers at the time in order to sleep) that there were no dreams that I registered in my mind during that period. Whether it's the cart before the horse I don't know." He dated the beginning of his recovery to the resumption of his dreams. The very night he left the hospital, he said, "It was as if my brain was announcing the return to well-being through this kind of amazing nonstop dream."

As Styron intuited, the entire pattern of the sleep cycle changes in depression. A healthy person typically falls into a non-REM stage of sleep that gives way to a REM period after about an hour and a half. This first REM phase lasts for around five to ten minutes, and each subsequent REM phase grows progressively longer. If you spend six to nine hours asleep, you have about four to six sleep cycles; by the last stage, you may be spending as much as an hour in REM. Among the depressed, the first REM stage comes too early and lasts too long; it may begin only forty-five minutes into the night and last up to twenty minutes. (The more severe the depression, the earlier it comes and the longer it can last.) Although the depressed spend a greater proportion of the night in REM sleep, the brain regions responsible for rational thought are overactive — which might prevent the brain from generating the distant connections necessary for dreaming. The emotional pattern of the night, according to one 1998 study, is also altered. Normally, dreams grow more pleasant with each REM stage, making us less likely to have to wrench ourselves out of mood-ruining nightmares in the morning. For the depressed, though, dreams follow the opposite course, starting off emotionless and becoming more distressing.

From the late 1970s through the mid-2000s, psychologist Rosalind Cartwright — whom Matthew Walker considers "as much a pioneer in dream research as Sigmund Freud" — carried out a series of studies on the dreams of a population with high rates of depression: new divorcés. For one experiment, she invited sixty people who were in the midst of a divorce — roughly half of them depressed — to spend three nights in the sleep lab on two occasions, once at the beginning of the divorce process and again twelve months later. At the start of the project, a third of the members of the depressed group reported dreams about their exes. By the end of the year, those who had dreamed about their partners at the outset were more likely to have recovered on both practical and psychological measures; their moods were more positive, their finances were more stable, even their love lives were more satisfying. Dreaming about the divorce, it seemed, had helped them get over it.

In another study, Cartwright took a closer look at the content of the divorcés' dream diaries, trying to pinpoint what made some dreams more therapeutic than others. This time, she tracked the dreams of twenty-nine women, nineteen of whom started off depressed, through the first five months of their separations. The ones who were on the road to recovery, she found, tended to interact with their dream-exes in a more active, assertive fashion. One woman saw her ex-husband embarrassing himself at a party and felt relieved not to be with him. Another expressed her resentment toward her ex and his new girlfriend. These dreams were vivid and convoluted; they featured a diverse cast of characters and drew together disparate strands of the dreamer's past and present. The dreams of the other group, meanwhile — the ones who would stay mired in depression — tended to be simple and unemotional, with the dreamer occupying a more passive role. In one characteristic dream, a woman stood by silently as her ex took his new love interest on a date. In another, a divorcée watched her ex look at a pair of shoes.

Dreams can help us cope, too, with universal life-cycle struggles, like coming to terms with death. The mourning process is messy and individual, but for most people, the work of grieving continues in sleep; in vivid, unforgettable dreams, the dead come back to us. In a 2014 study of nearly three hundred mourners at a hospice center in upstate New York, 58 percent could recall at least one dream about the person who had died. Although the dreams were not always pleasant, they usually provided some measure of comfort; they helped mourners accept their loss and led to heightened feelings of spirituality and an overall sense of well-being. Often, the dreams showed the dead person young and disease-free, savoring the pleasures of the afterlife or bearing some hopeful message for the living.

At some point during her long and happy marriage, Joan Didion got into the habit of sharing her dreams with her husband each morning. This practice wasn't about dream interpretation, per se, nor was it the product of some kind of deliberate incubation. It was an unburdening, an emotional dumping that helped her get on with her day. In fact, her husband, John Dunne, was only a reluctant participant in this ritual. "'Don't tell me your dream,' he would say when I woke up in the morning, but in the end he would listen," she later wrote.

Dunne died at the age of seventy-one, succumbing to a heart attack just after he and Didion sat down to dinner one night. He was telling her about the book he was reading, a new assessment of the causes of World War I, when he slumped over and stopped talking.

Finding herself suddenly lost without her husband of nearly forty years, Didion fell into a deep depression. Her grief was compounded by the ill health of her daughter, Quintana, who — still a young woman — had been comatose with pneumonia when John died. Irrationally, Didion blamed herself for the double tragedy. She felt like she should have been able, somehow, to save her husband, to cure her daughter. "Not only did I not believe that 'bad luck' had killed John and struck Quintana but in fact I believed precisely the opposite: I believed that I should have been able to prevent whatever happened," she wrote in *The Year of Magical Thinking,* her memoir chronicling the first year of her mourning.

For several months after John died, Joan didn't dream at all. It wasn't until her first summer alone that her dreams started to come back, and they usually starred John. In one of the earliest dreams of her widowhood, she and John were meant to fly from California to Hawaii on a group trip organized by Paramount. Joan boarded a plane at Santa Monica Airport, but she couldn't find John anywhere, so she got off and decided to wait for him in the car — and saw that the plane was leaving. She found herself alone on the runway. "My first thought in the dream is anger: John has boarded a plane without me," she wrote. She connected the feeling to her waking state. "Did I feel abandoned, left behind on the tarmac, did I feel anger at John for leaving me? Was it possible to feel anger and simultaneously to feel responsible?" She later recognized that dream as a turning point, after which she began to forgive herself for her imagined sins. "Only after the dream about being left on the tarmac at the Santa Monica Airport did it occur to me that there was a level on which I was not actually holding myself responsible."

Joan Didion's dreams hew to a pattern typical among the bereaved. A decrease in dream recall — or even the complete loss of dreaming, as she initially endured — is characteristic of the acute depression that often follows a sudden loss. As she began to recover from the shock of her husband's death, her dreams returned, helping her work through her pain.

After psychologist Patricia Garfield's father died, she decided to interview other women who had recently lost someone important and found that she could match their dreams to different phases of mourning. The nature of grief dreams changed as the mourner started coming to terms with the loss. At first, the departed seemed to come back to life, wanting to talk about the circumstances of their deaths. These "alive-again" dreams were disturbing, inflaming the survivor's irrational sense of guilt over "allowing" the person to pass away. Six weeks after his dad died, Philip Roth dreamed that his father returned to Earth, angry that he had been buried in the wrong outfit. "All that peered out from the shroud was the displeasure in his dead face," Roth wrote in his memoir *Patrimony*. The dreamer might feel resentful that the deceased has fooled him or caused him pain, or the dream might be pleasant in the moment but lead to a keen sense of loss upon waking. Dreams like these, though painful, can help the mourner understand that the deceased is really gone.

During the next phase, which Garfield called disorganization, the departed might reappear and say goodbye or set off on some obscure journey. One widower in a study of Garfield's dreamed of driving to the airport with his wife. When the couple arrived, she went on ahead of him, waved goodbye, and told him that he would join her later. The man interpreted this dream as permission to engage in life again and credited it with allowing him to reintegrate with the world and even to remarry. In the final stages — once the mourner has accepted the loss — she might experience pleasant dreams in which the deceased is young and well again or offers words of comfort or advice.

The dreams of one young woman Deirdre Barrett met, who had cared for her grandmother as she died of cancer, exemplify this cycle. Her earliest dreams reflect a psyche racked with guilt. In one, her grandmother said they needed to try her death over again — maybe this time, the girl could get it right. In another, she told her to call the police, because she didn't die of cancer; she had been poisoned. When the young woman was starting to feel better, she dreamed that she was a child again. Her grandmother gave her a warm bath, told her that she loved her, and explained that she was heading to heaven. "Ever since then," the woman said, "I have been at peace with my grandmother's death."

When grief is complicated — when the mourning process goes awry, and the survivor fixates on the loss — dreams are complicated too. According to Garfield, "deadly invitation" dreams — in which a dead friend or relative beckons the survivor to join him in the grave — can signify suicidal thoughts. After writer Daphne Merkin's mother died, she hoped that she might finally feel released from her never-ending disapproval. Instead, their twisted, obsessive relationship continued to play out in Merkin's dreams. "In the first years following my mother's death, the much-longed-for sense of liberation didn't come; instead she kept putting in an appearance in my dreams, many of them unsettling," she wrote in her memoir *This Close to Happy*. "In one of them she came equipped with a penis and the two of us made love; I remember waking up with a feeling of great joy, as though my long search for completion were over. Small wonder that I continued to feel as if a big hole had been made in my life, and thought on and off of trying to join her."

Dreams can also help us grapple with the universal trauma of contemplating our own mortality. The relationship between sleep and death is intimate, even uncomfortably close. Socrates entertained the idea that death itself is a dreamless sleep. According to one strand of rabbinic thought, sleep is equivalent to one-sixtieth of death; in sleep, the soul is said to separate briefly from the body, as it leaves permanently in death. In Greek mythology, Thanatos, personification of death, is the twin brother of the sleep god Hypnos. The dreams of the dying, those perched on the precipice of oblivion, have often been treated as sacred portals to the world to come. In nineteenth-century America, Andrew Burstein wrote, "people were always moved to record the dreams of the dangerously ill, half expecting some supernatural revelation." The last dreams of criminals on death row, reported to priests before their executions, were illustrated and reproduced for the pages of Victorian tabloids.

Across cultures, dreams intensify in the run-up to death, sometimes along predictable lines. Visitation dreams — in which, Kelly Bulkeley and Patricia Bulkley wrote in *Dreaming Beyond Death*, "a recently deceased loved one returns to provide guidance, reassurance, and/or warning" — have been described in cultures all over the world. In journey dreams, "traveling, passing, moving, changing locations, and crossing from one place to another . . . help the dying person anticipate what lies ahead," frequently inspiring "a fundamental shift from fearful despair to calm acceptance and even welcome expectation."

Nearly all of the fifty-nine hospice patients in a 2014 study had at least one vivid dream or vision in the weeks and months before they died, often featuring religious figures or family members who had already passed away. An eighty-one-year-old woman named Audrey said she knew she was ready to let go after five angels visited her in a dream. For Barry, eighty-eight, it was a dream about his mother that helped him release his grip on this world. He dreamed that he was driving, he didn't know where, and his mother — who had died years before — was comforting him, assuring him that she loved him and that everything would be okay. Patients described the majority of these visions as "comforting" or "extremely comforting," testifying that they eased the fear of death.

In 2014, historian Wojciech Owczarski interviewed one hundred residents of Polish retirement homes about their dreams. His subjects were contending not only with the ordinary stress of illness and old age, but with the shame of being abandoned. "In Poland, as in the majority of other countries of Central and Eastern Europe, staying in a nursing home usually involves a traumatic experience of isolation and rejection by the family," he wrote. "In Polish, nursing homes are commonly, if derisively, called 'old people's houses' and are perceived as 'places to die in.'" The Poles' dreams transported them back to their youth, reviving memories of happier and more exciting times. "They dream about what they lack and what they cannot experience while in a waking state," Owczarski wrote. But instead of producing nostalgia or regret, these dreams were a source of profound comfort and joy. One woman appreciated that in her dreams, she could talk to her daughter — who, in real life, had not visited her in a year. A dying man felt closer to his adult children after dreaming of happy scenes from their youth.

From the time he was a small child, Maurice Sendak was acutely conscious of the fact that he would one day die. His parents, Jewish immigrants from Europe, never let him forget about the Holocaust, and his artwork constantly circled themes of aging and loss. He dreamed up the creepy, beast-like characters of *Where the Wild Things Are* — his most famous creations — after hanging out with his own relatives at a shivah. Success did little to quell Sendak's existential terror. When an old friend from high school asked him how it felt to be famous, he answered morosely, "I still have to die." As powerful as his fear of death was his fear of snow: Sendak worried that his roof would cave in from its weight or that a storm would keep him from reaching the hospital in an emergency. And although he fostered certain juvenile habits — he loved cake and was exceptionally fond of his dog — he hated Christmas; as the childless son of Jews, he had always felt left out around the holidays. As he lay dying in a Connecticut hospital, Sendak had a vivid dream that incorporated both of his deepest fears: He saw his beloved nurse Lynn lying on a sofa before a giant painting of a small-town Christmas, the landscape covered in snow. But this, he confided to Lynn, was no nightmare; this was a beautiful, comforting dream.

A man named Bill, whose story appears in *Dreaming Beyond Death,* was devastated when his doctors told him that his cancer had spread and he would live for only a few more weeks. He fell into a depression; fear was the only emotion to penetrate his listless fog. A hospice minister went to visit him at home and found him wan and withdrawn. When she went back to see him a few days later, though, she "noticed a remarkable change in Bill's mood and vitality." This time, "his eyes were alive with interest" and "his expression was relaxed." All that had happened between the two meetings was that Bill — who had spent most of his life as a captain on a merchant marine ship — had had a dream. He was sailing through dark, uncharted waters and felt "the old sense of adventure" coming back; even though the sea was stormy and wide, he knew that he was on the right track. "And strangely enough, I'm not afraid to die anymore," he told the pastor. "In fact, I feel ready to go, more so every day."

For nonbelievers in particular, pop icons can take on the role of dream-saviors. In the 1990s, folklore scholar Kay Turner began collecting women's dreams about the singer Madonna. Many of the women she interviewed found emotional support in their Madonna dreams, waking with a sense of peace or resolution that persisted in their real lives. Twenty-nine-year-old Margie dreamed that Madonna summoned her into a bathtub, sang "Like a Prayer" while washing her with a sponge, and wrapped her in a fluffy towel. When she awoke, she felt that she "had been baptized by love and sex, that Madonna had somehow cleansed . . . [her] of past experiences with a violent ex-lover." Thirty-five-year-old Chris was living in a psychiatric hospital, struggling to heal from the scars of a long-ago assault, when

Madonna approached her in a dream and helped her design an awareness campaign about sexual abuse. The dream came at a time when Chris "was really going through a lot of emotional turmoil," she said, "so it felt like magic almost."

Ernest Hartmann, a psychiatry professor at Tufts University, discovered that the dreams of trauma victims often follow a predictable —and salutary—pattern. In the immediate aftermath, the survivor endures vivid nightmares that reflect the primary emotion of the ordeal, usually something like fear, guilt, or loss. ("A huge tidal wave is coming at me." "I let my children play by themselves and they get run over by a car." "I'm in this huge barren empty space. There are ashes strewn all about.") In some cases, these dreams directly replay the traumatic event; in others, they retain the overall mood but switch an element or two. As time goes on, the dreams start to blend images of the trauma with memories from the dreamer's past or stories the dreamer has read or heard. These dreams help put the trauma in perspective; they are a reminder that others have survived similar disasters and that the dreamer has lived through other insults and injuries. *You can assimilate this into the mental framework you already have,* the brain is saying to itself. *It's not so bad.* Once the victim has recovered, her dreams return to normal. One New Yorker who emerged from the subway on September 11, 2001, and looked up to see office workers leaping out of the Twin Towers was haunted for weeks by nightmares of the scene. As she began to heal, the content of her dreams changed. In one series, she gave the victims colorful parasols to ease their landing. (In PTSD, this system malfunctions; the dreams don't evolve but remain—like the victim's mind—stuck in the past. The dreams replay the traumatic event in an excruciating, unchanging loop without bringing in other memories or giving the dreamer a chance to take control.)

Alan Siegel, a psychologist in Berkeley with an interest in posttraumatic nightmares, recognized a research opportunity when his region was struck by two natural disasters in a row. In 1989, the Loma

Prieta earthquake caused sixty-three deaths and nearly four thousand injuries. Just two years later, a massive wildfire swept through the hills of Oakland, killing twenty-five and destroying thousands of homes. Shortly after the second disaster, Siegel recruited forty-two Californians, twenty-eight of whom had lost their homes in the fire and fourteen of whom had not, and asked them to start tracking their dreams.

When Siegel collected their dream journals at the end of the experiment, he found that the Northern Californians' dreams were far likelier than a control group's to feature themes of death, injury, and disaster. (To his surprise, people whose houses had withstood the fire seemed even more disturbed than those whose homes had been destroyed; 17 percent of the former group's dreams referred to death, compared to 11 percent of the latter group's and 5 percent of a control group's. Siegel hypothesized that their trauma was complicated by survivor's guilt.) Like the divorcés' early breakup dreams, which predicted their mental state months later, survivors' initial post-fire dreams indicated how well they would be faring at the one-year mark. If someone dreamed about the fire (or some other natural disaster — a tidal wave, a flood, an attack) and exerted a measure of control rather than passively observing it, she stood a better chance of recovering.

In another study, Canadian psychologist Kathryn Belicki gathered reports from twenty-eight women, half of whom had been sexually abused, in which they described their worst nightmares. When independent judges read through them, they were usually able to figure out whether or not a given dream came from someone who had been abused. Those dreams were distinctive: they featured a high incidence of violence, sexual scenes, threatening figures, and faceless men. The frequency of nightmares is another sign that we process trauma in dreams. In a separate study of more than five hundred undergraduate women, Belicki and a colleague found that women who had been assaulted suffered nearly twice as many nightmares.

Except in the case of PTSD, dreams almost never replay real

memories exactly — but they draw heavily from our waking lives, spinning the threads of personal experience into a tangled web of present and past. More than a century ago, Freud wrote that the dreaming mind churns up the "residue" of the day, and modern empirical studies suggest that half of dreams include at least an element of recent experience. They reflect ongoing activities as well as significant one-time events. Children in war zones have more violent dreams than kids in peaceful areas. Sports-studies majors have more athletic dreams than psychology students. When our environment changes, our dreams do too. Back in the late 1960s and 1970s — before pesky ethical guidelines made this kind of project less feasible — psychiatrist Howard Roffwarg had college students spend several days wearing special goggles that screened out green and blue wavelengths, making everything look reddish. Whether objects were naturally white, gray, or scarlet, they all assumed a pinkish tone. The effect was often disturbing. One student who had always loved hamburgers was so put off by the pink ketchup he saw through his goggles that he considered giving up meat altogether.

As the experiment wore on and the students adjusted to their new, reddish reality, their dreams began to change. At the outset, the red tone permeated about half of their earliest dreams; in later REM phases, the color spectrum went back to normal. By the end of the week, though, their dreams were suffused with a red hue all night long; about 80 percent of their initial dreams were abnormally rosy, and almost half were still red-tinted in the morning. Roffwarg's study suggested that at least one key facet of dreaming — visual perception — is in thrall to recent experience and that we register phenomenological shifts in our dreams as we adapt to them in daily life.

It is impossible, though, to predict exactly which experiences will ultimately make their way into our dreams. When I asked Robert Stickgold what type of memories are most likely to be incorporated into dreams, he ticked off three factors. "I would guess that it has to do with emotional intensity. It has to do with repetition. It has to do with recency," he said. But even Stickgold, who has been at the forefront of dream research for decades, had to guess. No one — not philosophers, not psychologists, not neuroscientists — can explain why a certain image strikes us on a given night or say for sure whether some long-lost friend or dead relative will ever resurface in our sleep.

But scientists have provided a few clues about the emotional and chronological factors at play. In the late 1980s, Canadian neuroscientist Tore Nielsen discovered a phenomenon he called the dream-lag effect. If an event of the day is going to turn up in a dream, it usually happens that same night, whether as a literal replay, an abstract representation, or a single element of scene or character. By the next night, the chance of that event appearing in a dream has fallen by half. If an event fails to appear in a dream the same night, it might turn up one week later.

In one study, Nielsen had volunteers watch a violent half-hour video of Indonesian villagers ritually killing and sacrificing water buffalo. The participants tracked their dreams every night, and the events of the film turned up around the same time in most of their dream logs: between one to three nights later, and then again on nights six and seven. That pattern might help us cope; it's possible that alternating intense nightmares with mundane dreams allows us to process an event without throwing off the normal sleep cycle.

Dreams do more than help people recover from trauma after the fact; they can also provide a measure of relief during difficult times. During the Civil War, lonely soldiers dreamed of their families and woke with a renewed will to survive. "Fascination — if not obsession — with dreams, spiritualism, and other religious experiences saturated the world of Civil War Americans," wrote historian Jonathan White, who scoured soldiers' letters for accounts of their dreams. "More than anything else, men in the field dreamed of those they loved back home . . . most soldiers saw romantic dreams as a welcome comfort."

Austrian psychiatrist Viktor Frankl, who spent three years imprisoned in Nazi concentration camps, wrote about how he and the other inmates — subsisting on rations of ten ounces of bread and a few cups of soup per day — dreamed of "cake, cigarettes and nice warm baths." In 2015, Owczarski rummaged through the archives of the Auschwitz Museum looking for psychologists' reports on survivors' wartime dreams. Reading through the transcripts, he noticed that the dreams often filled a therapeutic function, lifting the prisoners' moods, instilling a sense of hope, or shoring up their faith. The most frequent motif was what Owczarski came to call the caring dream, in which "the dreamer experiences care or support from his or her relative or some other figure, often a supernatural one," such as an assurance that he would survive Auschwitz. One prisoner dreamed of receiving a visit from a divine being robed in white and bearing a message of hope: "Don't worry, you will survive this hell." Even though the dreamer doubted the existence of God, he held on to this vision like a talisman. "This dream . . . was strongly present in my unconscious, and in tough moments in the camp I 'clung' to it as my only life saver." When another inmate was sick with typhus, he thought back to a dream in which he was being pulled, against his will, into a stormy river. He was on the verge of drowning when his dead brother approached, handed him a giant fish, and promised that he would be able to carry it. Other dreams transported the prisoners back home, letting them enjoy a brief respite from their waking hell and reminding them of the simple joys of normal life. "For a certain period of time, sometimes even several days, these 'freedom dreams' allowed us not to feel the nightmare of camp life so acutely," one survivor testified.

When the prisoners were set free, the character of their dreams changed. Romanian psychologist Ioana Cosman interviewed twenty-two Holocaust survivors and found a sharp contrast in their dream lives during and after the war. While they were in the camps, their dreams "presented . . . brighter and happier scenes." It was only after they had been released — physically if not mentally — that their dreams took on "a darker and horrific form," replaying gruesome scenes from the war or tormenting them with visions of family members who had been killed. Their dreams were adaptive, colluding in their self-preservation — postponing the nightmares until they were ready to confront their worst memories.

AN INTRIGUING NEW line of research suggests that it might be possible to tamper with memories while we dream — a finding that could have implications for how PTSD is managed. In 2015, a team of intrepid French neuroscientists led by Gaetan de Lavilléon set out to see if they could manipulate a mouse's path through a field by selectively stimulating its neurons while it slept, implanting a kind of artificial memory. They began by releasing mice into an open environment and tracking how their place cells fired as they entered different parts of the field. As de Lavilléon expected, the rodents ran around haphazardly, showing no preference for any particular area. When they fell asleep, their place cells would fire in a similar pattern, as their sleeping brains replayed their daytime journeys.

When certain place cells were activated, de Lavilléon would stimulate the sleeping animals' medial forebrain bundles — neural pathways associated with rewards, like sex and drugs. When the mice woke up and were once again let loose in the open field, their movements no longer seemed random. As if drawn by some invisible force, they gravitated toward the spots associated with the place cells that had fired while their reward centers were stimulated. Even though the mice had no rational reason to prefer those place fields over others, they lingered there four or five times longer. "The animal developed a goal-directed strategy for the [location], as if the animal had a conscious recollection that there was a reward there," Karim Benchenane, who helped conduct the study, told the *Scientist*. The study "suggests very strongly that new learning was actually possible during sleep, if you pair it with the right reward," neuroscientist Matt Wilson said. "They could essentially condition animals to prefer a particular location, not based on any experience they'd had, but just by reinforcing the location during sleep." This research is still incipient, but the potential applications are tantalizing; this was the first time that scientists had succeeded in implanting a conscious memory in a living thing as it slept, and it suggests a new frontier in trauma treatment.

If dreams of traumatic memories were paired with pleasurable rewards, those memories could be modified and lose their painful edge. "In principle, you could selectively change brain processing during sleep to soften memories or change their emotional content," Benchenane said.

Even though dreams about grief or anxiety may be painful, we should embrace them, knowing that they are helping us heal. If we want to make the most of the therapeutic potential of dreams, we must hone our dream recall; the more fully we can remember our dreams, the more insight we can wring from them. When we spend time parsing the meaning of our dreams, we end up confronting the very issues they refer to, speeding our emotional recovery. If we track them over time, we can draw hope from signs that we are on the mend, such as when a pattern changes or when we start to assert ourselves more. "Watch for any dream image that reflects improved protection and increased control," psychologist Patricia Garfield advised. "Turn it over in your thoughts and feelings, and absorb its power to heal."

Dreams play a critical role in regulating our emotions, in helping us process painful experiences, and in forming and even modifying memories. But like most potent forces, they also have a destructive side.

chapter 7

NIGHTMARES

I'M LATE TO WORK. I SLINK INTO THE OFFICE, KEEPing my eyes down, as if—by the kind of magical reasoning more often endorsed by toddlers playing peekaboo—my colleagues can't see me if I can't see them. The sales team is already on its feet for the buoyantly named standup meeting. I sidle into my seat and power on my computer without looking up.

I barely managed to drag myself out the door this morning; I woke with the memory of a nightmare flooding my brain. Last night—or, more accurately, in the later stages of REM this morning—I learned that I had a six-month-old baby but no recollection of carrying him or even of giving birth. Perturbed, I tried to find out how the baby had survived all those months without his mother's care; he looked scrawny and frail, more like a tiny, wizened old man, a Benjamin Button, than any normal infant. My editor appeared on the scene, cradling her healthy, rosy-cheeked baby in her arms. She bounced her son on her lap; he cooed and laughed. Grasping for direction, I tried to copy her moves, but they felt unnatural, and my pruney, grayish baby, wrapped in dirty blankets, knew I was faking it. He snarled at me, rolling his eyes, and I knew the baby would never forgive me or recover from my neglect. I came into a slow, reluctant consciousness, a gnawing sense of guilt hanging like a smog over the morning. I reached for the notebook on my night table and dutifully transcribed the dream. As I rode the train into Manhattan, the image of the abandoned child imposed itself over the usual subway scenes — bespectacled publishing types studying this week's *New Yorker,* a pair of talented teenage buskers dancing in the middle of the car.

I AM SUPPOSED TO BE writing a blog post about the racial politics of *The Nutcracker* — it's due tomorrow — but every time I look up from my computer and see my editor, I feel vaguely traumatized. She looks well rested, elegantly dressed as always, subtly made up, as though she has stepped out of the pages of some high-end lifestyle magazine; she squints at her screen, absorbed in her work. It is bizarre to consider that she has no idea of her star turn in my nightmare. I have rarely felt so profoundly out of sync with the world, so unsettled by how little we know of one another's inner lives.

I tilt my screen down and furtively — I hope — take my anxiety to the biggest repository of all anxieties: the internet. *Dreams about babies, dream pregnancy meaning;* I Google every variation I can think of. I'm so desperate for answers, I don't even care what my employer might think of my browser history. The results are predictably useless. A website called Dreamwell tells me that the baby might symbolize my inner child or a new project in my life. I rack my brain — do I have some new project that's at risk of going off the rails? The website Dream Moods suggests that a dream-baby "signifies innocence, warmth and new beginnings." That seems unlikely. I log onto my e-mail, but instead of asking the dancer playing Chinese Tea for his thoughts about cultural appropriation, I Gchat my best friend about my dream. "I have not recovered," I tell her. "I nearly took a pregnancy test this morning. Even though I am on my period." At noon, I am still obsessing over the memory; I Gchat the man who played the father in the dream. "It was definitely your baby," I inform him. I can practically hear the accusation dripping from my fingertips. "It was a devil baby."

"How are you?" my editor asks when I finally lift my eyes long enough to allow for a normal greeting. The truth is that I am exhausted, that I feel as though I haven't slept at all, as though I passed the night in hell, birthing a monster, instead of resting. There is no framework in which to talk about this. *I'm not feeling so great,* I imagine confiding, *I had this dream.* I do not imagine this going over very well.

Mildly bad dreams are a healthy preparation for real life, but not all bad dreams are therapeutic; nightmares are the glaring exception. Waking from a nightmare can leave people disoriented and panicked, unable to fall back to sleep; the fear of nightmares can deter people from even getting into bed, fueling a vicious cycle of insomnia. Yet for millions of people, they are a brutal fact of life. Nightmares can begin in kids as young as two and typically become more frequent throughout early childhood, reaching a peak around age ten. The majority of adults continue to suffer from nightmares from time to time; one study found that four out of five adults could recall at least one from the previous year. The most common nightmare scenario is pursuit (the dreamer is fleeing some malicious stranger, real-life foe, or supernatural monster), followed by attack (in which the dreamer is the victim of a violent assault). The *Diagnostic and Statistical Manual of Mental Disorders* estimates that about 6 percent of adults have at least one nightmare per month, and 1 to 2 percent suffer "frequent nightmares," which are twice as common among women. (Boys and girls experience nightmares at a similar rate, and the gender difference emerges between thirteen and sixteen, around the time that women overtake men in disorders such as anxiety and depression.) A bad dream can set the tone for the entire day, determining the dreamer's mood and coloring her perceptions of colleagues and friends.

Unfair as it may be, we wake up angry with the villains of our nightmares. In 2013, Dylan Selterman, a psychologist at the University of Maryland, designed a study looking at the impact of dreams on romantic relationships. For two weeks, his subjects — sixty-one people in long-term couples — wrote down their dreams and answered questions about their feelings toward their partners. Slights and betrayals perpetrated by dream-partners affected the actual relationship; if a woman had a dream about her boyfriend cheating, she was less likely to report feelings of intimacy the next day, and the couple was more likely to have a fight.

"Dream delusions" can be so intense that the dreamer wonders whether they were real. A married narcoleptic patient of Erin Wamsley's once dreamed about having an affair and felt guilty about it "until she chanced to meet the '[dream] lover' and realized they had not seen each other in years, and had not been romantically involved." Another woman Wamsley met started planning a family member's funeral before realizing that the relative had died only in her dream. For people who already struggle with the boundaries between sleep and waking life, dream-reality confusion can be a serious problem. "They wake up with these vivid memories and can't distinguish what actually happened," Robert Stickgold said. He told me about a narcoleptic Englishman in Australia whose boss called to tell him he had been fired. Without a job to keep him there, the man figured it was time to move home. "He was packing when he got a phone call from a coworker, who was just checking in because he hadn't been to work all week." He had only been fired in a dream.

When writer Julie Flygare was in college, she began to catch herself dozing off at awkward moments. An ambitious student at an Ivy League school, she was horrified the first time she realized she had fallen asleep in class; she ran to the bathroom and splashed cold water on her face, but the problem continued. She was too embarrassed to tell her professors, and her grades suffered. At Thanksgiving, she traveled for hours to visit her family, only to spend most of the holiday asleep on the couch. Once, she even felt her eyes flutter shut while she was driving. It was a relief when she finally got a diagnosis: narcolepsy.

The random bouts of uncontrollable fatigue were bad enough, but what Julie really came to dread were her dreams — dreams so vivid she began to lose track of what was real. When she looks back on her postcollege years, her lifelike nightmares serve as a sort of shorthand for the trajectory of her itinerant twenties. DC was the city where she grabbed a can of pepper spray and ran outside to confront a burglar — only to realize that he was a figment of her dream. In LA, she got mad at her boyfriend for stomping around the apartment, burping and yelling, on a night when he wasn't actually home. In Boston, she dreamed that a stranger had smashed a window and come inside to kill her. "I remember seeing his arms stretch out towards my neck," she said. "I remember shuddering in terror. I struggled and struggled. Then I looked up, and he was gone." She summoned the courage to get out of bed and track down the intruder. It wasn't until she saw that the windows were intact, her roommate sound asleep, that she realized she had dreamed the whole episode. "It felt like it had happened," she said. "I don't say, 'I had a dream.' I say, 'A burglar broke in,' because that was my experience."

It's been a decade since Julie's diagnosis, and she's learned not to trust her own memory. She tells herself that anything she thinks happened around bedtime probably didn't. But even after all these years, the emotional force of her dreams has never waned. She sometimes wakes up crying, confused. "As soon as I become conscious," she said, "I should know it didn't happen, but the transition between these boundaries is so fluid."

In severe cases, dream-reality confusion can even trigger mania. Turkish sleep researcher Mehmet Agargun collected studies of bipolar patients whose manic episodes were set off by nightmares. An eighteen-year-old high-school student Agargun called Mr. A. woke up one morning and told his father about his dream: the earth was trembling and people were running out of their houses, falling down and crying out. Mr. A. notified his family that doomsday was approaching, warned them to prepare for death, and was checked into the hospital.

WHILE DREAMING ABOUT stressful events can help us process them, replaying traumatic events is counterproductive. In the 1970s, psychologist Joseph De Koninck devised an experiment to see how dreaming about an upsetting experience would affect students' ability to cope with it. Just before going to bed, he had a group of undergraduates watch an old, violent public service announcement about workplace safety. Two factory workers lost pieces of their fingers in the machines—the camera lingered on their bloody stumps—and a third keeled over and died after a coworker accidentally shot a board into his chest. When one group of students fell into REM sleep, De Koninck played a particularly disturbing piece of the soundtrack in which the clumsy perpetrator of the last, fatal accident said gravely, the sound of a power saw rasping in the background, "I knew then that I, Lucky Williams, had killed a man, sure as though I had done it with my own hands, and it didn't have to happen." Meanwhile, another group slept in a quiet room.

As De Koninck predicted, the students who had been exposed to the soundtrack while they slept incorporated the film into their dreams at a higher rate. When everyone watched the video for a second time, the ones who had heard the soundtrack and dreamed about the film were even more stressed. Rather than softening the blow, the dreams had intensified it—just as nightmares can worsen the psychological impact of more serious traumas.

For people with posttraumatic stress disorder, nightmares that incorporate vivid flashbacks to the very events that victims want to forget can be crippling. "If the daytime is the occasional stalking ground of Freud's *daemons,* then the nighttime is their lair, an underworld of mystery and metamorphosis where they have free rein," PTSD sufferer David Morris wrote in *The Evil Hours.* "The daemonic night and its chief product, the nightmare, have always been a special hell for survivors." In the general population, nightmares are linked to self-harm. Adolescents in rural China who have frequent nightmares attempt suicide nearly three times as often as their peers. In a longitudinal study of more than thirty-six thousand Finnish adults, the rate of suicide was 105 percent higher among frequent than occasional nightmare sufferers.

ACCORDING TO THE American Indian Zuni, most diseases are a byproduct of bad dreams. The Rarámuri of northwestern Mexico believed that evil, pathogenic spirits lurked in dreams. The stress of nightmares, we now know, really can contribute to illness. "If your dream contains danger, the sympathetic nervous system will kick in just as it would if you saw real-life danger while awake," explained psychiatrist Jean Kim.

That stress response can trigger a flare-up of symptoms in patients with asthma or migraines. In 1996, psychologist Gail Heather-Greener asked thirty-seven migraineurs — people who regularly suffered from migraines — to track their dreams until they had noted five that preceded an early-morning migraine and five that did not. The dreams that came before the headaches featured a higher proportion of anger, violence, and fear and were more likely to insert the dreamers in situations beyond their control. The causes of migraines aren't fully understood, but both psychological and physical factors play a role; stress — and the resulting drop in serotonin — can be a powerful factor. The migraineurs' nightmares might have provoked the stress response, or their dreams could have been a reflection of stressful situations in their real lives. But whether they're a symptom or a cause, the nightmares appear to be an aspect of the migraine trigger.

My friend Katy has been struggling with migraines for eight years, ever since her final exams in high school. "When I first got them I thought I was going mad," she said. "It felt like there was bubble wrap around my brain and mind." She's gotten better at managing her symptoms, but they still interfere with her life sometimes. She gave up coffee and has been told to stop drinking alcohol. "When they are bad I can't move and have to lie in a dark room with a pillow over my head," she said. "Sometimes, my brain just stops working and it feels like wires are fusing." Before a migraine begins, Katy often has intense nightmares. "There's definitely a link for me. The worst one was probably that my brother died, and I kept trying to talk to him even though he was dead. I also had one where I was on trial for murder, and another one where I was suspected of smuggling drugs at an airport. I woke up really stressed out and dizzy, which is a sign of my migraines." Another migraineur I know keeps a diary to monitor potential triggers — including the pollen count, her food intake, and her dreams. "Nightmares often come hand in hand with a migraine," she said. "I'm trapped in a tower or a bunker, or there's an environmental disaster — a fire, flood, or hurricane — and my family can't escape."

Physical pain is typically present in less than 1 percent of dreams, but that number rises among the sick. When doctors in Montreal asked burn victims to record their dreams through the first week of their hospital stays, 39 percent reported at least one pain dream. One man dreamed "that there was fire burning inside of me." Another dreamed that grenades, disguised as bowling balls, were falling from the ceiling. These dreams can interfere with recovery; patients who reported pain dreams slept more fitfully, felt more stress when they thought about their injuries, and experienced more acute pain during their treatment. The pain dreams could have been a reflection of their daytime state, but the doctors believed that they contributed to "a cycle of pain-anxiety-sleeplessness" that hobbled the healing process.

The Western medical literature even includes cases of seemingly healthy people succumbing to heart attacks within hours of waking from vivid nightmares. A man in his late thirties — a nonsmoker with no family history of heart disease — dreamed that he died in a car crash and woke up vomiting; two hours later, he was at the hospital describing the unbearable pressure in his chest. A twenty-three-year-old woke at six a.m. from a nightmare in which he was murdered alongside his father and had a heart attack at seven. The early-morning and the final hours of sleep — when REM cycles are longest and nightmares are most intense — is the most dangerous period for cardiovascular patients; heart attacks are most frequent, and most severe, between the hours of six a.m. and noon.

Over the course of a few years in the 1980s, seemingly healthy young men in the Midwest began dying in their sleep, one by one, baffling doctors and epidemiologists. They passed away in the early hours of the morning, lying on their backs, with looks of horror in their eyes — 117 in total. Doctors examined their diets, their heart function, their psychiatric histories, all to no avail. "We did an autopsy in each case, and we got a big zero," a bewildered medical examiner told the *New York Times*. The men's unusual background offered the only clue. They were members of an isolated hill tribe known as the Hmong, and they had recently emigrated from Laos to the United States, mostly to Minneapolis–St. Paul and California; they had spent, on average, just under a year and a half in the United States.

Back in Laos, they had helped the Americans fight the North Vietnamese, and the combat had been savage. By the time Laos lost the war, in 1975, almost one-fourth of the Laotian Hmong population had been wiped out. The survivors fled, and tens of thousands sought asylum with their former ally. America took them in, but only grudgingly. Immigration officers sent them off to different cities, spreading them "like a thin layer of butter throughout the country so they'd disappear," as one resettlement officer put it.

The Hmong immigrants had escaped a literal war zone, but starting anew in a foreign culture came with its own challenges. They struggled to learn English and find steady work; their unemployment rate was as high as 90 percent. They couldn't maintain long-standing generational hierarchies and gender roles. Parents swallowed their pride and asked their children to translate English, and men had no choice but to accept whatever money their wives could bring in. Depression was rampant, and suicide — unheard of back in Laos — became an issue. "I want to die right here so I won't see my future," a middle-aged Hmong man — a former soldier who was living on welfare in San Diego — told a journalist.

Shelley Adler, an anthropologist at the University of California, San Francisco, spent fifteen years studying Hmong religion and interviewing Hmong who had lived through the crisis. In 2011, she published her findings: The cause of that fatal epidemic was "the Nightmare" — or at least, the victims' belief in it. Deprived of their livelihood and the tight-knit communities they had always counted on, they became so afraid of an evil nightmare spirit they called the *dab tsog* that their hearts stopped beating in their sleep. No one can fully explain the physical mechanism that killed the men, but Adler argued for a combination of stress, biology, and sheer terror.

In Hmong theology, the *dab tsog* is thought to target her victims while they sleep. When she appears in women's dreams, she causes infertility or miscarriage; when she comes for men, she is deadly. Her first strike is rarely fatal; her targets have time to appease her by consulting a shaman or offering sacrifices. It isn't until the third attack that the *dab tsog* aims to kill. She is particularly vicious toward people who have neglected the demands of ancestor worship — elaborate rituals that were impossible to perform in American cities, far from the sacred Laotian mountains and the ancient family graves. About half of the immigrants who attempted to practice their native religion in the new country received a visit from *dab tsog,* but the episodes were even more common among Hmong who converted to Christianity — perhaps a consequence of their guilt.

Even the men who survived a *dab tsog* attack could be traumatized. Victims described the feeling of a heavy weight clamping down on their chests and the terrifying sensation of being unable to move. Their wives — helpless and equally desperate — looked on as the men convulsed and turned blue. One fifty-eight-year-old survivor, Chue Lor, told Adler about his harrowing encounter with the *dab tsog.* "I was in my bed at night. There were people at the other end of the house and I could hear them talking . . . I heard everything. But I knew that someone else was there. Suddenly there comes a huge body — like a big stuffed animal they sell here. It was over me — on my body — and I had to fight my way out of that. I couldn't move — I couldn't talk at all. I couldn't even yell, 'No!' . . . I was trying to fight myself against that and it was very, very, very scary."

Adler does not deny the role of biology in the men's deaths; ECGs of some of the men who were hospitalized revealed that they were prone to ventricular arrhythmias. But psychological stress can be a powerful trigger for cardiac arrest. The men's day-to-day lives were already a battle, and the lethal *dab tsog* episodes often followed something especially tense, like arguing with their families or hearing bad news. Sleep debt could also have exacerbated the attacks; often, the victims had stayed up late the night before, watching TV or — in ironic attempts to stave off the *dab tsog* — trying not to sleep. When the men finally, inevitably passed out, their brains would compensate for their exhaustion by sending them straight into intense REM periods — fertile territory for vivid nightmares.

THERE IS NO failsafe way to tame nightmares. A drug called prazosin, which works by blocking the effects of norepinephrine, an adrenaline-like chemical that can contribute to nightmares, is sometimes prescribed to help PTSD patients sleep. But it's unreliable; patients risk side effects like nausea and headaches, and some even have trouble breathing or pass out the first time they take it. Imagery rehearsal therapy (IRT) — drawn from the arsenal of cognitive behavioral techniques — can help, but it's time-consuming; patients have to spend at least ten minutes a day replaying their nightmares and visualizing new endings. And the process is so unpleasant that few stick with it. For PTSD patients in particular, rehashing traumatic nightmares can do more harm than good.

One rainy Monday morning in May, I decided to try out the latest technology in nightmare treatment for myself. I took the train to Boston, where Patrick McNamara was at work on a program he hoped would one day cure bad dreams. In a cramped, carpeted room at the Center for Mind and Culture in Kenmore Square, I donned a black headset that completely enveloped the upper half of my face, jutting out a few inches into the room. In a process based on imagery rehearsal therapy, I, like the other subjects in McNamara's pilot study, would practice manipulating nightmarish images. Unlike in traditional IRT, though, I wouldn't have to summon the images myself; the virtual-reality headset Oculus Rift would do it for me.

For years, scientists have been experimenting with virtual reality as a tool in treating phobias and PTSD. Someone with a fear of flying might sit in a chair that vibrates like a plane in turbulence while watching a screen that simulates the view from the sky. A patient trying to overcome a fear of public speaking might be dropped onto a computer-generated stage before a crowd of virtual spectators. "The rationale is that by confronting the trauma in a safe environment and talking about it and recasting it in the safety of the clinical office, eventually the anxiety will extinguish," said psychiatrist Skip Rizzo, who studies the clinical uses of virtual reality. "One of the reasons that PTSD tends to become chronic is that people try to avoid thinking about the trauma. When you avoid something you fear, you get a temporary sense of relief, and that relief reinforces continued avoidance. We're trying to circumvent that avoidance by helping a patient go back and relive the scene in a safe environment."

The images in McNamara's study were chosen to evoke the same psychological reactions as nightmares. Each one coded highly on one of three emotional dimensions: valence, arousal, or dominance. "If it's a high negative valence, then it would be highly unpleasant for the viewer to experience," explained Kendra Moore, a research assistant who was helping to run the study. "If it has a high arousal score, it's very arousing for the viewer to look at. And if it's highly dominating, then it would be dominating over the viewer." Patients would use the hand controls to manipulate the images and render them less threatening, shrinking the scary parts or painting over them and making up stories to explain the transformation. McNamara expected that the amount of time patients spent manipulating the images would correlate with a reduction in nightmares and improvements at a test of visual-imagery control. In this test, subjects were asked to call to mind a scene — in one case, a car parked in a driveway — and then imagine various changes: See the car in a different color. See the car driving down the road. See the car crashing into a house. People who struggle with this type of mental acrobatics also struggle to control the images that make nightmares so nightmarish. "Many of the people that are coming in and doing this treatment have an overactive amygdala, which is where all the fear regulation happens," Moore said. "They have less control of the imagery process in their prefrontal cortex." The treatment "is practice for people to gain better control over those brain processes that contribute to nightmares."

As Moore explained the study, she used little spheres of black plastic to draw a three-foot circle around me; she was marking off the circumference of the world I was about to enter. I lowered a heavy black mask over my eyes and quickly forgot how silly I must have looked; I forgot about the outside world entirely. I had entered a magical hotel lobby, and as I did a slow 360-degree turn, I found new spectacles wherever I looked. Flames of bright orange cubes rose from a virtual fireplace, emitting a hypnotic crackling sound. An urn, a statue, and a red robot sat side by side, as though the vestibule had been decorated by aliens let loose at a garage sale. Outside, cherry trees — whose pink blossoms looked like they had been crafted out of paper lanterns — billowed in a preternaturally blue sky. Whenever I approached the edge of the "play space" Moore had outlined, a shining turquoise grid materialized out of thin air.

Once Moore had tired of watching me watch the inside of my goggles, she pressed a button, and the lobby faded away. A new image appeared in front of my eyes as though it had been projected onto the wall of some space-age classroom: a litter of puppies. It was a control, designed not to provoke much of a response. I clicked, and the dogs disappeared. Next, I saw a butterfly perching on a flower; another neutral. After I had swiped through the warm-ups, the real program began. I saw a close-up of two shiny black cockroaches, their spindly antennae casting creepy shadows, the hairs on their legs glistening in high relief. Moore couldn't have chosen an image more likely to elicit a negative reaction if she had psychoanalyzed me for years. (In some instinctual version of IRT, I wrote my fourth-grade research paper on cockroaches, but my fear persisted; ten years later, living on my own for the first time, I fled my apartment — and reconciled with an ex — after seeing a water bug on the bathroom floor.) I froze when the image popped up, and as soon as I recovered the capacity for movement, I frantically clicked with my right pointer, which had taken on the faculties of a paintbrush, and used it to draw black pixels over the image. I held my finger down until the face of one bug morphed into a nebulous blob and I felt my pulse start to return to normal. I kept blotting out the roaches' bodies until they merged into a nondescript patch of black; I added a pair of crude ears and decided that it was a sheep. The image had lost its shock value, reduced to a nonthreatening — if misshapen — farm animal.

I clicked through to the next image, bracing myself — but this one, which I later learned counted as high-arousal, didn't have much of an effect. A group of skydivers were poised, their arms spread, about to jump out above a tree-dotted landscape. Their protective gear looked effective; I didn't find the image especially ominous or even arousing, so I scrolled on without altering it. Next, I saw a close-up of a snake, poised as though ready to strike, taking up the entire screen; this one was high-dominance. I painted over its long, forked tongue — which was sticking out, as though the snake were preparing to swallow the viewer — and then shrank the whole thing by moving my left hand from side to side. Scaled down, the reptile instantly lost its power.

The exercise was over, and I pulled the headset off. The lab seemed drabber than I remembered, the textures somehow flat, the sky outside the window disappointingly dull. I would have happily come back for more sessions, more opportunities to visit this alternative reality. Which is the point: Virtual-reality therapy is much more popular than traditional IRT. In one study, 114 out of 150 phobic patients said they would rather be treated with VR than IRT. "When you do traditional imaginal exposure therapy, you're asking the patient to close their eyes and imagine what they've been practicing avoiding," Rizzo said. "That's a pretty tall order, and you never know what's going on in the imagination. Virtual reality is a more systematic, controlled, potent way of doing this exposure."

Another effective way to deal with nightmares is much less high-tech. If people can learn to become conscious in their dreams, they can wake themselves up or even banish their dream-foes. In 2006, psychologists at Utrecht University in the Netherlands set up a study to explore whether lucid-dreaming treatment could help people overcome their nightmares. They recruited twenty-three men and women whose bad dreams frightened them awake at least once a week and split them into three groups. The people in the first group received a single, one-on-one session of lucid-dreaming therapy in which they learned about lucid-dream induction techniques and were told to conduct a reality test whenever they felt afraid or recognized a situation that reminded them of a nightmare. They practiced re-scripting their dreams, making up new endings in which they wrested control back from their demons and averted the usual horrors. Members of the second group went through the same steps but learned the techniques in small groups rather than in individual sessions. The last group — the control — was relegated to a waitlist and received no treatment at all.

The first two groups went home with instructions to keep practicing, and when they returned to the lab twelve weeks later, the interventions seemed to have been a success. At the start of the experiment, volunteers in group therapy had an average of 3.1 nightmares per week, but by the end, they had, on average, only 2.6 nightmares per week;

patients who learned in private sessions started out with an average of 3.6 nightmares, and that number went down to 1.4, while the control group, which started out at an average 3.7 nightmares a week, held steady at 3.6. The improvement didn't depend on achieving lucidity; several people who never managed to become lucid in their dreams still had a reduction in nightmares. Practicing the re-scripting exercise provided some good in itself.

For twenty years, writer Steve Volk was plagued by the same terrifying dream. Every few months, usually when he was sick or stressed, he would dream of a strange man hovering outside his window, threatening to come in and kill him. After several gut-wrenching minutes, the stranger would break into the house and start beating him up. Volk grew accustomed to waking up in a panic, his hands balled into fists. He learned about lucid dreaming while doing research for a book on fringe science and decided to see if LaBerge's techniques could help him. He called the Lucidity Institute and an instructor advised him to play out the nightmare while he was awake, pinpointing the moment at which he wanted to become aware. Volk picked the instant when the bully's face appeared at his window, and he pictured it over and over.

One night, he felt the dream coming on, but this time "I was there," he told a radio interviewer in 2012. "My perspective shifts and I am in this body, in this place — not observing something but in it. I could feel my fingers tickling my palms. I could feel my feet on the floor . . . I go to the door, I reach for the door, and the door handle is a door handle — it feels that real." The man comes in, and there's a twist; for the first time, his attacker is armed. "The dream becomes a battle between what I know to be true — that this is a dream, and it has no external reality — and the natural feelings of fear that crop up when somebody who's been terrorizing you for twenty years pulls out a gun," Volk recalled. But when the man started shooting, Volk realized that the bullets were not hurting him. It was only a dream. "I woke up feeling like Superman," he said. The nightmare never came back.

Unchecked nightmares can wreak havoc on our waking lives, even on our physical well-being. But promising new treatments are in the works, and the stress of nightmares can trigger lucidity; becoming lucid—as we'll learn more about later on—can turn nightmares into liberating adventures. And true nightmares make up a small minority of our dream lives—most dreams, even bad ones, are salutary.

chapter 8

DIAGNOSIS

A USEFUL SIDE EFFECT OF THE THERAPEUTIC function of dreams is that, if we pay attention, we can see what our brains are trying to process. And if we take dreams out of our journals and bring them into doctors' or therapists' offices, they can become a valuable diagnostic tool. This is what some scientists would call a spandrel — a term borrowed from architects, who use it to describe the triangular space that forms when two arches are built beside each other. Spandrels are an accident — they came into being as a mere byproduct of the useful parts of an aqueduct — but Roman artists recognized the opportunity they presented and began carving intricate designs and religious symbols into them. "They started to do fancy artwork on them, but they're there to serve a different function," dream researcher Robert Stickgold said. In biology, a spandrel is something that evolved as a byproduct of something else but for which people have invented a function — like the sound of the heart beating. "Your heartbeat is there because your heart is a pump," Stickgold said. "But we have found great uses for it." Physicians can detect murmurs by listening to the heart beat, but your heart doesn't beat to alert your doctor to problems. Likewise, Stickgold argued, dreaming is "something that just happens to happen, and so we take what use we can of it. People have learned to exploit dreams . . . I think dreams are as useful to a creative person or to an introspective, psychologically minded person as your heartbeat is to your physician."

The very first use that psychologists identified for dreams — the grounds for Freud's fixation — was their ability to show us our flawed, neurotic selves. Contemporary psychologists, after trying for years to shed their association with Freud, are finally coming back around to that forgotten truth: Dreams have an important role to play in diagnosis. They can expose anxieties we don't intend to divulge and fantasies we didn't know we harbored. Whether in knotty symbols that must be untangled through analysis or in literal depictions of real-life scenarios, dreams reveal the messy hearts of our emotional lives.

One of the obstacles to treatment is that therapy is predicated on honesty: on confessing weird symptoms and self-destructive habits, dredging up distant memories and old traumas. The expensive, time-consuming irony is that patients lie. In a survey of more than five hundred people undergoing therapy, almost everyone — an astonishing 93 percent — admitted to having lied in session. (They most often hid suicidal thoughts, drug use, and disappointment in the therapeutic process.) No matter how much patients trust therapists to suspend judgment, to keep their secrets, to take their side, they can't help but lie to elicit the response that they want; to elude censure or punishment; to save face or shield the doctor from discomfort.

Dreams are invaluable in diagnostic terms because they let patients off the hook. It can be easier to admit to something that happened in a dream, which can always be sourced to the murky pits of the unconscious or even blamed on a random physical stimulus, than to broach an embarrassing fear or voice an irrational anxiety. Some languages, like Greek, implicitly recognize this distance between dream and design, allowing a speaker to deny responsibility for a dream and claim the passive role of witness rather than the active one of author. "I saw a dream," a speaker of Greek might say, rather than "I dreamed."

In many religious traditions, people are automatically absolved of sins they commit in dreams; the dream-self can't be held accountable for succumbing to temptation. "What a man does while he sleeps and is deprived of reason's judgment is not imputed to him as a sin, as neither are the actions of a maniac or an imbecile," wrote Thomas Aquinas on the subject of "nocturnal pollution." (Not all spiritual authorities are so forgiving. Many male Orthodox Jews and Kabbalists would traditionally undergo purification after having a wet dream. On Yom Kippur, the high priest would historically force himself to stay up all night to avoid dreaming of sex.)

Ever since Freud, a group of psychologists have continued to treat dreams as the royal road to the unconscious. Psychoanalyst Stephen Grosz, for one, has elicited breakthroughs by probing his patients about their dreams. In *The Examined Life,* Grosz recalled how one patient — a sixty-six-year-old widow he called Elizabeth — would come to every appointment and complain about some minor crisis: she lost her wallet and keys; she spilled a glass of red wine on a friend's couch; she forgot her sister's birthday lunch. Although Elizabeth existed in a state of perpetual anxiety, she and Grosz kept running out of time before they could talk about bigger issues, like her husband's recent death. For months, this charade went on. "There was always some new problem that required urgent attention," Grosz wrote. She could never remember any dreams.

Only after a year of analysis did Elizabeth begin to open up about her husband's final months, confessing that, as he lay dying of pancreatic cancer, she withdrew, avoiding him and taking any excuse to leave the house. Desperate to escape the reality of his looming death, terrified by his neediness, she turned away.

Around this time, Elizabeth finally told Grosz about a dream. The telephone rang at home, and she knew it was her husband calling, but she couldn't find the receiver; it wasn't in its usual place. The phone's endless ringing taunted her as she tore up the house, searching for it, to no avail. She woke up crying, and she wept again as she recalled the dream — the first time she had broken down about her husband in therapy. The dream — and her outburst in remembering it — finally allowed Grosz to crack through her emotional armor, to understand the litany of minor disasters and the immense guilt brimming beneath the surface. "There are various ways to circumvent depressed, anxious feelings," Grosz explained. "It's not uncommon, for example, to exploit sexual fantasies, or to use hypochondriacal worries. Elizabeth employed her disasters to calm herself — they were her tranquilizer . . . we can sometimes exploit a disaster to block internal change."

Many contemporary psychologists are unsympathetic to Freud, but a growing body of empirical research suggests that issues we try to ignore during the day really do bubble up in our dreams. People who frequently suppress negative feelings — who endorse statements like "I always try to put problems out of mind" and "There are things I prefer not to think about" — are more likely to dream about emotionally fraught memories. Even if they manage to avoid their problems in the day, they can't escape them at night.

Psychologists have a term for this: the *dream-rebound effect*. By the time Daniel Wegner turned his attention to the question of whether dreams churn up topics we would rather avoid, the Harvard social psychologist had already determined that trying to suppress thoughts was, in general, a fool's errand. In the 1980s, he asked a group of undergraduates to narrate all of their thoughts for five minutes, with one constraint: They were not allowed to think about white bears. Whenever a student said or thought the words *white bear,* he had to announce the failure by ringing a bell. Wegner's subjects — college students in a major American city who had little reason to be obsessing over wild animals native to the Arctic Circle — rang their bells at least once a minute.

In the next phase of the experiment, the students were allowed to think about white bears as much as they wanted. This time, they rang their bells even more frequently and — crucially — even more often than a different group of students, who had been encouraged from the start to contemplate white bears. Not only had the attempt to repress the image failed; it had led to a rebound effect, a period in which they could think of little else. If we make a conscious effort to avoid a thought, we have to focus on that effort—which makes the whole project impossible. The rebound effect has since been replicated in all sorts of real-world situations. Smokers who try to avoid thoughts of cigarettes end up obsessing over them; dieters who strive to suppress images of chocolate can picture little else; depressed people who try too hard to think positive instead ruminate on worst-case scenarios.

Wegner knew that during REM sleep, parts of the brain involved in thought control shut down, and he wondered whether those thoughts that people tried to quash in the day—those white bears—would pop up in dreams. To test this idea, he assigned more than three hundred students to pick someone they knew and then spend five minutes before bed transcribing their thoughts. The first group was told to avoid thinking about the person they'd chosen; a second group was instructed to concentrate on that person; and a third group was told to give a quick first thought to the chosen person and then think about whatever they wanted. In the morning, they all wrote down their dreams, and the effect was clear: the first group produced more dreams than any other about the target person.

JUST AS FREUD argued, dreams may be especially adept at revealing what we desire. South African scholar Mark Solms is at the forefront of neuropsychoanalysis, a small field that uses brain science to reconsider Freudian ideas. When he discovered that patients with damage to the pontine brain stem were still able to dream, he realized that Hobson and McCarley's activation-synthesis theory (which proposed that dreams were triggered by neurotransmitters in the pons) didn't tell the whole story.

Hobson and McCarley didn't know that dreaming isn't confined to REM sleep; it's also possible to dream as we fall asleep and right before we wake up. In all of those scenarios, the brain is aroused in some way

—by the flow of acetylcholine in REM, by the dregs of daytime consciousness at the beginning of the night, by the release of hormones before waking. If the pontine brain stem wasn't providing the primary driving force for dreams, then what was?

Neuroscientists talk about a handful of emotional-processing systems in the brain that respond to fear, panic, rage, and seeking—basic systems that we share with animals and that involve some of the most ancient areas of the human brain. They are set off by external stimuli—the sight of a snake, for instance, might activate the fear system and trigger an internal response, like shifts in blood flow and increases in heart rate.

The seeking system, also known as the reward system, energizes us to take an interest in our surroundings and explore our environment; it also motivates the appetite for food, water, and sex. When a need-detector system registers an imbalance, it stimulates appetitive behavior—if our thirst detectors pick up on low levels of water, for instance, we're motivated to find something to drink.

But during dreams, muscles are paralyzed and the body is still. Asleep and immobile, we can't go looking for the rewards we're programmed to seek; we end up imagining them instead. More evidence of this came from a study in which sleeping subjects were injected with dopamine stimulants, which activate the seeking system. They enjoyed "a massive increase in the frequency, vivacity, emotional intensity and bizarreness of dreaming."

DREAMS CHANGE IN distinctive ways as people cope with emotional problems and even psychiatric disorders. Once we learn to understand the language of our own dreams, then—whether we share them with a therapist or keep them to ourselves—we can recognize when something changes. Although there is an amazing range in how different people dream, there is a surprising constancy in each individual's dreams over the course of her life; we each express our fears and fixations in our own vocabulary, returning to the same symbols and characters over the years. The themes and style of our dreams usually hold steady from youth through old age — the proportions of men to women, of relatives to strangers, of friendly to hostile encounters; the number of animals, the frequency of sex. "Cultural stereotypes portray dreams as irregular and infinitely varied, but as the findings on repeated themes and repeated elements in long dream series clearly establish, dreams are in fact extremely regular and repetitive," University of California, Santa Cruz, psychologist William Domhoff wrote.

In *The Individual and His Dreams,* Calvin Hall and Vernon Nordby described a long dream sequence from a woman they called Dorothea who started recording her dreams in 1912, when she was twenty-five, and kept up a detailed dream diary for half a century. Many themes remained constant over the years. In one out of six dreams, she lost some possession. In every tenth dream, her mother appeared. Every sixteenth, she fretted over missing a bus or a train. But other themes shifted in tandem with her changing anxieties and station in society. In middle age — but not when she was young and busy — Dorothea dreamed of feeling left out. Those dreams tapered off as she got older and accepted her social life and her status as a single woman.

Recurrent dreams are particularly easy to track. It's estimated that between 60 and 75 percent of adults have had recurrent dreams, usually triggered by stress. The majority are unpleasant, if not downright nightmarish, and the most common one involves being chased — adults, according to Canadian psychologist Antonio Zadra, by "burglars, strangers, mobs and shadowy figures"; children by "monsters, wild animals, witches or ghoulish creatures." People who are experiencing recurrent dreams typically have elevated scores on measures of depression and complain of more problems in their lives. The cessation of a recurrent dream is a sign that the underlying issue or source of stress has been resolved. In one study, people who had had recurrent dreams in the past but no longer did had better mental health than those who had never had them. It was as if they "had been forced to exercise their mental muscles more strenuously to overcome some deficit and were consequently more fit than those who hadn't been challenged to exercise so vigorously."

Kelsey Osgood still remembers the food-centric anxiety dreams she endured as an anorexic teenager decades ago. "One was about cereal, and one was about frozen yogurt," she recalled. "I would be eating a lot of it, and when I woke up, I would be unsure about whether I had actually eaten it. That was very distressing for me." Around the time she was hospitalized for her anorexia, she started having nightmares about being trapped in a supermarket. "I used to grocery shop a lot. Well, it wasn't really shopping, because I wouldn't buy anything — I'd get confused and come out with nothing. I used to have anxiety dreams about being stuck in large grocery stores with food everywhere and not being able to find the cash register."

Kelsey has been healthy for the past five years. She no longer dreams about gorging on artificial health foods or losing her way in mazelike supermarkets, but she still has the occasional anxiety dream about her size. In one recent dream, a journalist called her fat. "Whatever publication it was offered me a million dollars for emotional damages, but I was so upset that I refused the money," she said. When she wakes from a dream like that, she can still be rattled. "I feel annoyed that it comes out in that vocabulary. I like to think of myself as being fully above that. It upsets me that it is there, even in the smallest of subconscious ways."

The mere presence of food in dreams can be a sign of an eating disorder. In one study, half of bulimic patients and one-quarter of anorexics dreamed about food on a single night they spent in a sleep lab. (When, in my early twenties — overwhelmed by the need to forge my own path after leaving the cocoons of high school and college — I sought some kind of structure by obsessively tracking and then cutting my food intake, I dreamed of feasting on ice cream, four-year-old asparagus, and once, memorably, a platter of disembodied human nipples.) In healthy people, meanwhile, dreams about eating are relatively rare; food was mentioned in only about 1 percent of the dreams in Calvin Hall and Robert Van de Castle's original sample from Case

Western Reserve University, and even scenes of cooking and restaurants appeared in just 16 percent. A more recent study of Canadian college students confirmed that dreams are a culinary desert — less than one-third could recall ever enjoying a meal in a dream.

Just as anorexics dream about the substance that obsesses them, the newly sober dream about their own preoccupation; 80 to 90 percent of addicts in the first weeks of abstinence experience vivid dreams about drinking, using, buying, smoking, sniffing, or somehow interacting with their drug of choice.

Sarah Hepola first blacked out from beer when she was eleven, and by the time she graduated from college, she had a full-blown drinking problem. In her memoir *Blackout,* she recounted binging on hard liquor and losing entire nights, sleeping with strangers and stripping in front of crowds. In her mid-twenties, she got sober, and new interests took the place of booze. She saved money, read copiously, traveled around South America. It was a period of growth and exploration, and for over a year, she managed to resist the ever-present temptation. But she still missed her wild nights out, and she sometimes fell off the wagon in her sleep. Her dreams took on a feverish, inevitable quality. "I'd be at a party, and somebody would hand me a drink. Right as I'd start drinking, I'd remember I'd quit," she recalled. In the morning, she would feel as though she'd been cheated. "I remember waking up and being like, 'Crap, I can't drink.' It was like the fun had just started in the dream, and then I'd wake up and be like, 'Oh no, you're back in this terrible reality where you're not allowed to drink anymore.' It felt like I was doing all the hard work of sobriety, but I was haunted in my dreams. The more I was having these dreams, the more angry I was getting."

It was — among other things — a particularly vivid drinking dream that led to her undoing after a year and a half of sobriety. The details of that fateful dream remained etched in her mind nearly two decades later. She was at a baseball game, and someone handed her a cup. She took a sip and realized it was beer — but instead of spitting it out, she took another sip, and then another. "I woke up from that dream and I very consciously decided, 'If I'm going to have these dreams, I might as well drink.' Then I drank for ten years."

Hepola didn't know that drinking dreams are as common in early sobriety as sleepless nights and the shakes. Addiction counselors warn patients to expect them; drinking dreams are so predictable that psychiatrists use them to gauge the risk of a relapse. The dreams typically taper off as cravings subside, so a sudden drinking or drug dream after a period of abstinence can be a warning sign. The addict's emotional reaction in the morning provides the most important clue. If she dreams about getting wasted and wakes up feeling guilty, that may bode well for her recovery; even a pleasant dream about drinking or taking drugs is a good omen if the patient is relieved to realize that it was just a dream. A friend of mine who went to rehab in her early twenties fits into this theory. "When I quit, I dreamed about it for a year and actually felt drunk in dreams," she told me. "The drinking dreams I had were usually lousy and I woke up like, 'Phew.'" She's been happily sober for ten years.

Waking up feeling jilted or disappointed, however — as Hepola did — may be a sign of waning resolve. People are especially likely to wake with their willpower depleted if their efforts to procure a drug were thwarted in their dreams — perhaps the police showed up or the needle kept missing the vein. Dreams like this remind the user of what she's missing, triggering cravings without satisfying them even in the dream state.

Dreams can also help predict whether someone is on the cusp of a manic or self-destructive episode. (One of the most famous breakdowns in Western literature is precipitated by nightmares: "As Gregor Samsa awoke one morning from uneasy dreams, he found himself transformed in his bed into a gigantic insect.") In the 1990s, Canadian researchers Kathleen Beauchemin and Peter Hays spent six months tracking the dreams of people with bipolar disorder. Three mornings per week, they would phone each of their patients and inquire about their mental state and their dreams of the previous night. When their moods were stable, their dreams centered on mundane, realistic activities like working or commuting. But when they were manic, they featured "magic, delusions, aliens, flying, unusual creatures and experiences out of the ordinary."

Most usefully, Hays and Beauchemin were able to find patterns in the dreams that immediately preceded a depressive or manic episode. Depressive periods were typically prefaced by an overall drop in dream recall, while mania would be heralded by a night or two of vivid dreams about injury, violence, or death. One patient, right before a manic phase, dreamed of being lowered into a coffin and watching her family grieve. Another dreamed of walking through a cemetery and seeing corpses rise out of their graves.

One of the most urgent issues psychiatrists face involves an alarming degree of guesswork. There is no foolproof way to predict who will attempt suicide. Scientists have experimented lately with blood tests and app-based algorithms—with iffy results—but most have overlooked a more intuitive source.

Myron Glucksman, a psychiatrist in Connecticut, is an exception. For the past few years, he has been combing through depressed patients' dream reports, and he's managed to identify certain themes common to both the suicidal and nonsuicidal depressed as well as a few key differences. Both groups' dreams dwell on death and feelings of hopelessness, but the dreams of people who have contemplated or attempted suicide are more likely to feature destruction, aloneness, and self-directed violence. Shortly before she was hospitalized for suicidal thoughts, one patient of Glucksman's dreamed that her father ordered a man to shoot her. Two weeks before trying to poison himself with carbon monoxide, a man dreamed that an atomic bomb explosion "sucked the life out of people."

In 2017, Glucksman visited a local psych ward and asked fifty-two people with severe clinical depression to describe a dream from the previous two weeks. He divided the patients into three cohorts: those who had been diagnosed with depression but not suicidal ideation (Group A); those who were contemplating suicide (Group B); and those who had made a serious suicide attempt within the past fourteen days (Group C). He and his coauthor, Milton Kramer, then read through the dream reports, scanning for recurrent images and themes.

Dreams from Group A featured an unusual amount of loss, failure, and despondency. (For example: "I was in a snake pit with snakes slithering around me. I couldn't get out and I felt helpless, hopeless, and frightened.") The dreams of Groups B and C were also marked by loss and failure, but they were set apart by a fixation on death, violence, and murder. ("A little kid was stabbed in the chest and I saw it. I was the little kid and saw myself dying"; "I was with a group of bikers and there was a shooting . . . I was shot and was dying.")

"That makes sense because suicide is self-murder," said Glucksman. This method "could be used by clinicians — both mental-health clinicians and emergency room department physicians and people who work in crisis-intervention centers." It doesn't even require interpretation or analysis: "All you have to do is ask the person, 'Have you dreamt?' And if in the dream there's imagery of violence and murder and injury and you suspect the person is clinically depressed and may be suicidal, that is a predictive factor. I think it can save lives if people pay attention to this."

In his own practice, Glucksman uses dreams to measure his clients' progress. "There are different variables we can track over time," he said. "The stories that are reflected in the manifest content of their dreams change. The relationships change. Their self-representation changes." In *Dreaming: An Opportunity for Change,* he described how the evolution of one depressed patient's dreams convinced him that she needed to be hospitalized. In a recurring nightmare, the woman would find herself treading water in the ocean, clinging to a rock, and would see Glucksman passing by in a rowboat. He would extend a hand, but she could never quite reach him. The dream came back night after night, with a disturbing twist: Over time, the woman's grip on the rocks grew looser and looser. When she admitted that she had completely lost her hold, Glucksman sent her to the hospital.

Even without professional therapists, people can use dreams to recognize their own problems. A nightmare prompted Carrie Arnold, a science writer who chronicled her struggle with anorexia in *Running on Empty,* to realize that her eating habits weren't healthy. Not long after she was hospitalized, she dreamed that she let herself eat as much lettuce as she wanted. "I woke up and I was literally drooling," she said. "My pillow was sopping wet."

Carrie's first reaction when she woke was not embarrassment about the saliva drenching her sheets but groggy horror at the thought that she might have actually binged on salad. It was only when she remembered that she was locked up in a hospital, without access to a kitchen, that she calmed down. Even in the depths of her denial, it struck her that this dream — this nightmare — was not exactly normal. "I remember thinking it was so yummy and it was so exciting to let myself eat as much as I wanted of anything," she said. "But there was also the sense of realizing that this was bizarre, that lettuce was the only food I would even let myself dream about. I remember thinking it was kind of sad. That was as much insight as I was capable of at that point."

Kelsey Osgood told me about an anorexic friend whose nightmare finally helped her recognize the depth of her illness. Though her friend was dangerously sick, she resisted treatment; she believed that she should be allowed to eat as little as she pleased. In the dream, she married a skeleton. "If you read that in a novel, you might think it was a little obvious," Kelsey said. "But the dream made a big impact on her. She came in to group therapy more moved than I had seen her. That dream was able to rattle her in a way that being in a hospital hadn't."

Even the most rational people can't always acknowledge their own vices. For years, the renowned sleep researcher William Dement smoked up to two packs of cigarettes a day, doing his best not to think about the impact on his health — until he was diagnosed with lung cancer in a dream. "I remember as though it were yesterday looking at the ominous shadow in my chest x-ray and realizing that the entire right lung was infiltrated," he recalled. He was overcome by the excruciating knowledge that he was dying, that he wouldn't live to see his children grow up. "I will never forget the surprise, joy and exquisite relief of waking up." He quit smoking that day.

DREAMS — IF WE SHARE THEM — can even help doctors catch physical problems; as Aristotle and Hippocrates suspected, dreams often change over the course of a disease. The incubation period for illness and fever is often marked by heightened dream recall and distinctive nightmares. The first recorded use of the term *fever dream* dates to 1834, when the English writer Felicia Dorothea Hemans began her poem "The English Martyrs" with the narrator — a prisoner — sarcastically greeting the new day: "Morn once again! Morn in the lone dim cell / The cavern of the prisoner's fever dream." In 2016, German sleep researcher Michael Schredl asked a group of young adults to report the most recent fever dream they could remember. When he compared these to a sample of dreams from healthy students, he found the fever dreams to be much more bizarre, often featuring some kind of spatial distortion, like burning clouds, moving walls, or menacing blobs. People rated their own fever dreams as more intense and all-around negative; more than a third qualified as nightmares, with many incorporating real-life symptoms.

"Fever can be associated with delirium, a dangerous condition where the brain can hallucinate and consciousness waxes and wanes due to some underlying medical-physiological toxicity (often related to toxins from infection or drug reactions or other medical causes)," said Jean Kim, a psychiatrist at George Washington University. "Although we probably don't know enough yet to assign very specific dream content to specific medical conditions . . . dreams are probably a portal to a mind-body link that we need to continue to explore." In dream sleep, explained Patrick McNamara, "You're doing enormous sensory processing without regulation, without being inhibited. It makes physiological sense that if you're going to pick up faint signals from the body, like if you have an internal organ that's malfunctioning, that would be the time."

In the mid-twentieth century, a Russian psychiatrist named Vasily Kasatkin collected more than sixteen hundred dream reports from patients at the hospital where he worked. He analyzed the development of their symptoms alongside the dreams and picked out a handful of relevant patterns. Just before someone fell ill, he was often afflicted by "dreams with a very unpleasant and even nightmarish character," incorporating scenes of "war, fire, injury or other damage to different parts of the body . . . blood, flesh . . . dirt, dirty water . . . the hospital, pharmacy, doctors, medicines, etc." Sometimes, nightmares were more specific; one patient dreamed about his abdomen — the site of his future ulcer — being chewed by rats. The emotional tenor of these dreams matched their gruesome content, with patients recalling a crushing sense of fear or sadness in the morning. "Often, unpleasant dreams appeared before other overt clinical symptoms of the disease," Kasatkin wrote. He hoped that they could one day serve as an early warning system. "If doctors constantly monitored dreaming of patients throughout a disease and the period of recovery, they would often, if not always, notice changes in the nature of dreams."

Sometimes, dreams reflect physical symptoms in a way that doesn't require constant monitoring or even much analysis. When Oliver Sacks was a young doctor, he decided to spend his summer vacation hiking in Norway — a long-awaited break from his high-pressure job at a New York hospital. One August morning, the physician set off before dawn, alone, to climb a steep mountain, six thousand feet above a fjord. "I soon got into my stride — a supple swinging stride, which covers ground fast," he recalled in his memoir *A Leg to Stand On*. He was making steady progress, relishing his solitude, admiring the scenery ("a dark and piney wood . . . a new fern, a moss, a lichen"), when his meditations were disturbed by the sight of a "stupendous" bull blocking his way. Terrified, he bolted down the mountain — "madly, blindly, down the steep, muddy, slippery path" — and tripped and fell down a cliff, badly injuring his left leg. For seven hours, he dragged his broken leg down the mountain, crawling and sliding, wondering if he would survive, before he was finally rescued by a pair of reindeer hunters.

The experience was traumatic, but his recovery was even more disorienting. Sacks-the-doctor had become Sacks-the-patient, and though surgeons succeeded in reattaching his torn quadriceps muscles to his knee, something still seemed wrong. He felt alienated from his injured body. "I had lost the inner image, or representation, of the leg," he wrote. "Part of the 'inner photograph' of me was missing." Nowhere was this more undeniable than in his dreams. He recalled one especially disturbing nightmare: "I am on the mountain again, struggling to move my leg and stand up . . . the leg was sewn up, I could see the row of tiny neat stitches. I think, 'I'm ready to go!' But the leg wouldn't budge . . . not so much as the stirring of a single muscle fiber. I felt the muscle — it felt soft and pulpy, without tone or life . . . The leg lay motionless and inert, as if dead." In fact, in spite of the doctors' insistence that he was fine, Sacks was suffering from denervation (a loss of nerve supply) in his leg, which doctors finally recognized a few months after the dreams.

This wasn't the first time Sacks noticed a link between dreams and early symptoms. As a young neurologist, he was often struck by the continuity between the zigzag patterns that troubled migraine sufferers during the day and the images they saw in their dreams. In his book *Awakenings,* he recalled a man who was haunted by dreams of being frozen before he was diagnosed with Parkinson's disease.

Modern scientists are only beginning to untangle the intimate relationship between sleep and illness, but some progressive doctors believe that dreams really do incorporate symptoms that might otherwise go unnoticed. Narcoleptic patients, in one study, had an uncommon number of dreams about being paralyzed. Four of five patients in one sleep-disorder clinic who dreamed about sweating complained of excessive perspiration in real life, and nearly half of patients who were choked in their nightmares had trouble breathing in the daytime.

Dreams might help doctors catch difficult-to-diagnose disorders like sleep apnea, in which people briefly stop breathing in their sleep. Patients often don't notice their symptoms, which present while they're unconscious and leave no physical traces. One of the few signs is a long period of unpleasant or emotionally shallow dreams. In a 2011 study, doctors at Britain's Swansea University asked forty-seven patients at a clinic for sleep-disordered breathing to write down their dreams for ten days. Patients whose sleep apnea was more acute—whose breathing stopped more often or for longer intervals—tended to have flat, affectless dreams, devoid of the usual range of dizzying imagery and psychological intensity. The emotional poverty of their dreams might be a consequence of frequent interruptions in their sleep; it "could be that the sleep of patients . . . is so fragmented that it interferes with the process of dreaming, thus not allowing dream plots and dream emotion to develop."

One of the most useful applications of dream-related behavior is in determining who is at risk for neurodegenerative disease. Middle-aged people who act out their dreams—walking around or talking in their sleep—are significantly more likely to develop Alzheimer's or Parkinson's later on. So-called REM sleep behavior disorder (RBD) is most common among men over fifty and can turn otherwise peaceful people into menaces, dangers to themselves and their bedmates. The medical literature includes stories of RBD patients breaking their bones, chipping their teeth, punching their partners, and even leaping out of windows. A few years ago, psychiatrist Carlos Schenck caught up with twenty-six men who had been living with RBD for at least sixteen years and found that an astonishing 81 percent of them had been diagnosed with Parkinson's or a similar neurodegenerative disorder in the interim. (The diagnosis followed the RBD one by anywhere from five to twenty-nine years.) Other studies suggest that a diagnosis of RBD means a 50 percent chance of developing Parkinson's or dementia within the decade. Scientists don't fully understand the link between RBD and cognitive decline, but there are a few physiological parallels, deficiencies that can present as RBD before progressing to something more serious. Both sets of patients suffer from low levels of dopamine and an impaired sense of smell; both have lesions and distinct cellular abnormalities, called Lewy bodies and Lewy neurites, in the brain stem.

Even after a diagnosis has been confirmed, predicting how an illness will unfold involves another round of guesswork — and, once again, dreams can offer clues. For a year and a half, Jungian analyst Robert Bosnak led a monthly dream group for heart-transplant recipients. As the patients recovered from surgery, they struggled to conceptualize the foreign organs as real parts of themselves and even to overcome an irrational survivor's guilt; they felt lucky and grateful for their second chance but also estranged from their bodies. These psychological conflicts often surfaced in their dreams. One woman dreamed of being stalked by hooded, knife-wielding phantoms taunting her and telling her she didn't deserve to live. Another dreamed about walking through a wall and woke with the eerie sensation that she was no longer fully human; she felt as though a ghost had taken over her body. Dreams of metaphorically accepting the heart, however, were linked to recovery. When one woman started to feel better, she dreamed that her donor presented her with a beautiful red rose.

ALTHOUGH SCIENTISTS ARE starting to draw links between dreams and diagnosis, the world that Galen and Hippocrates envisioned — the world in which physicians routinely ask patients about their dream lives — has not materialized. Even so, people who pay attention to their own dreams can glean a powerful, if not always explicable, awareness of their bodies.

In 2011, Rebecca Fenwick (whom I connected with through the Facebook page of *Lucid Dreaming* author Robert Waggoner) was eagerly anticipating the arrival of her third child. Eight weeks into the pregnancy, she and her husband went to the doctor for a routine checkup and saw their baby's heart beat for the first time. She went home happy, but that night, Rebecca — who had always paid attention to her dreams — had one of the most frightening lucid dreams of her life. She was standing in her kitchen when she noticed a tornado forming outside the window — a typical sign that she might be in a dream. She did a reality test and, sure enough, realized that she was asleep. "There are levels of awareness within lucid dreaming, and in this one I was very, very much aware of what was going on," she recalled. Right away, the dream took a dark turn. "I was watching myself and also living myself in the dream, like a dual consciousness. I could see myself in the kitchen, with a glass of water, and — it's embarrassing, but I started to bleed. I grabbed a paper towel and wiped. I realized, within the dream, that I was having a miscarriage." Rattled, she managed to yank herself out of the nightmare. She woke her husband to tell him about the dream, but he brushed it off. "He's not really into that sort of thing," Rebecca said. She wanted to agree with him, and when morning came, she tried to put the dream out of her mind and went outside for a stroll. *Everything feels fine,* she remembered thinking. *Everything feels normal.* "There was nothing to make me think that I was having a miscarriage.

"When I got back from the walk — it was probably a two-mile walk — I went into the kitchen and I went to get a glass of water, and I started drinking the water. I was walking through the kitchen and I felt a gush. I thought, *Oh my God, maybe I peed my pants*. I went into the bathroom and I wiped, and sure enough, I was bleeding heavily." Even before she and her husband began to grapple with their disappointment, the first wave of sadness was tempered by déjà vu. "I was distraught and sad and emotional," Rebecca said of the moment she realized she had miscarried. "But I think I was more shocked that the dream was real."

As much as we can learn about our bodies and minds by tracking our dreams and perhaps discussing them with doctors, we can learn even more if we share them more widely.

chapter 9

DREAM GROUPS

ONE AFTERNOON IN 2016, I SAT IN A THERAPIST'S office in Manhattan and told him about a dream. And six of my friends were listening.

When I first met Mark Blechner, he raved about how much he had learned about himself when he submitted one of his dreams for group analysis. "I had never seen a dream so thoroughly worked on and understood," he told me of the dream group he took part in at a meeting of the International Association for the Study of Dreams in the 1990s. "Seven strangers made a dream much more clear than I had ever experienced." Mark was immediately likable; he exuded an air of quiet authority, and it was easy to imagine opening up to him. Even his office seemed to emanate nonjudgmental orderliness. Floor-to-ceiling bookshelves lined the wall. Porcelain figurines and an American Indian pot—a gift from a former student—sat atop tomes with FREUD 1, FREUD 2, and FREUD 3 printed on the spines.

When he got home from that IASD conference in Hawaii, Mark, who teaches psychology at New York University, began leading dream groups for his own students, excited to expose them to this alternative method of dream analysis. Several of the dream groups he set up have continued to meet without him. "It's extraordinary how much you can learn about dreams through dream groups," he said. "Every time I present one of my own dreams, I am flabbergasted. It gives me profound insights into myself."

I was intrigued. I had never heard of a dream group, but I wanted profound insights into myself, and Mark said that if I could round up several friends, he would try to help me find some.

I asked Mark if he had any guidelines for the dream we would analyze. He assured me there was no ideal type or length for the dream; he could work with whatever I brought. I scanned my dream journal, looking for an entry that would provide enough material to sustain a two-hour analysis without revealing any especially embarrassing secrets. I settled on a dream I'd had about a month earlier.

> I am walking down the street when Hillary Clinton invites me into a line dance. It's comfortable, as if she's an old family friend. The scene changes, and we're sitting at my mom's kitchen counter and talking about presidential pets. I have to convince her not to choose Serena, my step-mother's dying cat, as First Cat. We are talking about cats and dogs when Hillary suddenly slumps over and stops talking, and I realize that she's dead. I call the Secret Service, but they don't seem that concerned; I can't make anyone grasp the gravity of the situation. I run to get Chelsea, but she's just annoyed that I am interrupting her. Finally, I realize that Hillary is actually just a small slice of avocado.

It seemed like a safe choice: a bizarre episode that had left me wondering what psychic dramas or real experiences could have inspired it but that, I was pretty sure, couldn't possibly be interpreted as some kind of Oedipal conflict or draw out any awkward confessions.

I WAS WRONG, of course.

One Tuesday that March, I coaxed a group of friends from Brooklyn to Mark's office on the Upper West Side.

Throughout the session, Mark explained, I—"the dreamer"—should feel comfortable, in control; I could refuse to answer any question, and we would move on to the next phase only when I agreed it was time. I handed out copies of the dream to my friends and, per Mark's instructions, read it aloud. I felt my cheeks turn hot; the dream sounded even more ridiculous in this clinical setting. It was as though I had wandered into some kind of absurd writing workshop where, instead of fielding critiques of my word choice or plot structure, I answered questions about the "manifest content" of my dream. How did Hillary's hair look? (Short, 2016-campaigning-style.) Was there anyone else in my mom's house? (No.)

When, with everyone's help, I had thoroughly excavated my memory, we progressed to the next stage. Each member of the group free-associated with the dream, taking it on as her own, imagining that she had personally experienced it and offering her own associations—connections between the dream and her waking life, feelings the dream would have evoked for her. During this stage, I was not allowed to participate; I stayed silent as my friends pretended that they had dreamed my dream. "If the group members start without the dreamer's associations, their responses will be free and spontaneous, coming from where they are in relation to the dream," explained Montague Ullman, who devised this method of dream analysis. "The responses may sometimes be wide of the mark, but they will often be on target—and those responses might have been screened out if the group members had been working along lines suggested by the dreamer."

"If it were my dream, I might associate Hillary Clinton with my mother, who is a big Hillary supporter," one friend volunteered. "If it were my dream, I might see Hillary as an opportunist—using the line dance as a way to pull me in and convince me to vote for her," said a friend who preferred Bernie Sanders. "I can just imagine her trying to act natural in a line dance."

"If it were my dream, I might feel that I had been abandoned by someone who was supposed to protect me," offered another friend, who noted the lack of authority figures in general—the unresponsive bodyguards, the unconscious Hillary, the empty house.

Next, I answered questions about links between the dream and my real life. By this point, the drama—the mystery of what this dream meant to the person who had actually dreamed it—had heightened; after spending half an hour guessing what it would have meant to them, they wanted to know what it meant to me. They asked about my feelings toward Hillary Clinton (favorable) and Serena the cat (hostile). We discussed the linguistic link between *avocado* and *avocat,* the French word for "lawyer"—was that why my subconscious had turned Hillary, a lawyer, into an avocado? Moira pointed out the resemblance between an avocado and a vulva. Mark noticed the absence of men.

We talked about the time I'd found Serena frozen in a seizure and wondered whether she was dead. We decided that I feared being unable to communicate something important—that Hillary Clinton or the country was in danger, or some story or truth in my work as a writer. Mark then assigned one of my friends to read the dream back to me, this time at a comically slow pace; as she read it, I thought about what everyone had said. Finally, the group reflected on everyone's contributions and incorporated the ones I deemed relevant into a new interpretation. We had all shared personal stories about politics, about parents, about childhood homes, and I left not only feeling closer to everyone I had analyzed the dream with; I also left with a greater understanding of the dream and the real anxieties that might have inspired it. A whole group of thoughtful laypeople can do more thorough work than a single therapist. Different members of the group pick up on different layers of meaning—some are apt to notice echoes of family dramas; others tend to focus on political anxieties or gender dynamics. And committing to discuss dreams with a group can motivate people to remember them. "A primary reason why most adults in industrial societies do not remember their dreams is that they have no socially and emotionally supportive context in which to do so," wrote Jeremy Taylor. "My experience of more than twenty years is that as soon as such a context is created (say, for example, by their taking a class in dream psychology, or joining an ongoing dream group), then even the most stubborn failure to recall dreams is usually overcome."

One of the saddest consequences of our cultural contempt for dreaming is the trope that dreams make for boring conversation. In a society that still sees dreams as frivolous, airing them aloud is considered pointless at best, self-indulgent at worst. People worry that in sharing their dreams, they could inadvertently reveal some shameful neurosis or deviant desire; one of Freud's most enduring—yet least supported—theories is that most dreams express unconscious erotic wishes. If someone says, "You were in my dream last night," it's still basically an innuendo.

When Shane McCorristine, a scholar of modern British history, went trawling through police reports from nineteenth-century England, he was struck by the number that contained descriptions of dreams: witnesses and victims seemed to make a point of telling police and coroners if they had anticipated a crime or a death in their dreams. Telling dreams, he said, was a way to create "a social bond between a vulnerable person and the authorities." But he noticed that dream reports started dropping out of inquests and news stories in the 1920s, and he pinned the blame on Freud. "Freudian theories were spreading, and they were recalibrating people's relationship with the dream world," he said. "There's increasing embarrassment around dreams." Suddenly, they might be interpreted as signs of some latent neurosis or sexual deviance.

A century later, conventional wisdom dictates that dreams are not a subject for polite conversation. Writing for the *New Yorker*'s website in 2018, Dan Piepenbring began a review of *Insomniac Dreams*—the book about Nabokov's dream experiment—by apologizing for the topic: "Dreams are boring. On the list of tedious conversation topics, they fall somewhere between the five-day forecast and golf." ("My editor never challenged my description of dreams as boring," Piepenbring told me.) A few years earlier, radio producer Sarah Koenig devoted an episode of *This American Life* to laying out the seven topics that interesting people should never talk about. Dreams came in at number four, right behind menstruation. In the *Guardian,* British writer Charlie Brooker claimed that listening to other people's dreams made him dream "of a future in which the anecdote has finished and their face has stopped talking and their body's gone away." Novelist Michael Chabon wrote in the *New York Review of Books* that discussion of dreams is all but banned from his breakfast table, railing against them as poor conversational fodder: They drag on and on. They get twisted in the telling. Most unforgivable, they are bad stories. When I explain the topic of my book, people frequently offer their sympathies: "People must want to tell you their dreams," they say with an I-feel-your-pain nod. "Those are the most boring conversations."

"Tellers of dreams have some basic obstacles to overcome," literary scholar James Phelan said when I asked him whether there was anything about dreams that rendered them tedious narratives. "What makes stories of non-dream experiences interesting is that they are 'tellable' in some sense: the story implicitly claims that there's something about the experiences that raise them above the level of ordinary, unremarkable happenings." The protagonist might confront some danger, learn a lesson, or encounter something beautiful. But in dreams, "just about any event can occur, which means that the ordinary/extraordinary distinction relevant to stories of non-dream experiences no longer applies, which makes tellability more murky." Another problem is that dreams don't follow the type of logic we expect of a good yarn, Phelan said. "Often tellers will try to recount faithfully the sequence of the dream events. But such faithfulness typically means no cause-and-effect logic, and that absence typically means no coherence to the story, and no coherence means a bad story.

If the story of my day is boring because it is awash in details of no significance, the faithful recounting of a dream is boring because it is awash in randomness."

And it's hard to feel invested in another person's dream. You don't have any stake in it—you know from the outset that the story ends with the dreamer waking up in bed, unscathed. "The teller of the dream has a listener who inherently doesn't really care, because it's the teller's dream, and the listener is hearing something kind of egotistical and likely to be embarrassing," said Alison Booth, an English professor at the University of Virginia who specializes in narrative theory. "How are we to imagine we are the dreamer, when we hear about it? Whereas in fiction, rule number one is you are the reader and you have every right to be at the center of the story/imagine yourself as protagonist."

But maybe Westerners are just out of practice; maybe they don't know how to communicate their dreams. The reluctance to talk about dreaming is a culturally specific—and recent—phenomenon. There may even be an evolutionary reason why we feel so compelled to share our dreams. If the brain is trying to identify weak associations that may be valuable, then "it's got to be very lenient," Robert Stickgold said. "Maybe part of this process of biasing the brain's association-strengthening mechanism—to say, 'Pay attention to this association I found'—carries over into waking, and now you want everyone else to pay attention to it."

As our ancestors intuited, talking about dreams—whether casually recounting them to friends, analyzing them in structured groups, or even sharing them with strangers on the internet—can amplify their benefits. The more we integrate our dreams into our days, the more easily we remember them. And the act of discussing dreams can bring people together; just as dreams open up conversations on sensitive or embarrassing issues in a therapeutic setting, they can also facilitate intimate conversations among friends. Social psychologist James Pennebaker, who has spent decades studying the psychological effects of keeping secrets, has found that divulging hardships can make them feel more manageable. Victims of abuse are more likely to suffer all kinds of physical and psychological ills if they don't disclose their traumas, as psychologist Meg Jay noted in her book about childhood adversity. Secrecy leaves them susceptible to everything "from ulcers, flu, and headaches to cancer and high blood pressure," Jay wrote. According to Pennebaker, "the act of not discussing or confiding the event with another may be more damaging than having experienced the event per se."

Dream groups and the disclosures they inspire can form the backbone of communities. The formula that Mark Blechner follows, I learned, is based on Montague Ullman's dream-group model. After spending much of the 1960s and 1970s running experiments on dream telepathy at Maimonides Medical Center, Ullman grew restless. "I experienced a sense of diminishing returns," he wrote. He felt overwhelmed by the administrative duties of directing a lab, and he was itching to try something new. He left the hospital and took a temporary teaching job in an idyllic Swedish city to rest and contemplate his next move. He started talking to his colleagues at the Psychoanalytic Society about how they could appreciate their own dreams, which constituted "a radical shift in focus from their patients' dreams to their own. But my enthusiasm must have been contagious, for they began to respond to the power of the process we were engaged in." After his teaching job ended, he signed up for workshops at Esalen and other "growth centers," and by the time he returned home, he had found his new mission. He wanted to democratize dream analysis — to find a way for people without special qualifications or access to psychiatric care to gain insight and social connection from their dreams. "Trust, communion, and a sense of solidarity develop rapidly in a dream-sharing group," he wrote. "There is an interweaving of lives at so profound a level that the feeling of interconnectedness becomes a palpable reality."

Until his death in 2008, Ullman traveled the world promoting dreams as a self-help tool, leading dream groups, and teaching others how to set them up. His followers carry on his teachings with something resembling a religious zeal. "He's not here to train and inspire us," wrote William Stimson, who leads dream groups in Taiwan and maintains a website devoted to Ullman's legacy. "It's up to us to devise various ways to be inspired by and receive training from one [an]other. We are what he has left behind."

New research confirms what Ullman suspected: participating in a dream group can yield a host of social and psychological benefits. Like Mark Blechner, Mark Blagrove learned about dream groups at the annual IASD conference. The mild-mannered English psychologist turned up at his first meeting with no expectations and hesitated to share his own dream; it was "so short and obviously meaningless." In it, his partner, Julia, gave him two CDs that were stamped with a portrait of Rembrandt in a floppy hat, and he wondered whether Rembrandt was the guy who wrote the theme song for *Friends*.

But when he started discussing it with the group, Mark was staggered. The epiphanies came one after another. Julia had given him two CDs — just like she had given him two children in real life. The present was artistic — and she had sacrificed her own career as an artist to support his academic ambitions. The way Mark described Rembrandt's floppy hat reminded one participant of a scholar's cap; in fact, Mark had just been promoted to department head and full professor. By the end of the session, he was drawing connections between Julia's contributions and his professional achievements and feeling a renewed sense of gratitude for his family. "It was a tiny little dream that was not very interesting, and yet it had a whole load to it," he said. "I just wanted to keep going after that."

He started leading dream groups at Swansea University, and he could hardly believe the personal stories and problems his students were willing to share — their struggles adjusting to college, their relationships with their parents, their feelings about leaving home. "It's incredible, the lack of inhibition they've got," he said. Mark has devoted much of his energy over the past few years to exploring the psychological impact of belonging to a dream group. In one study, he and his colleagues measured students' levels of personal insight after they shared either a dream or a significant real-life experience with the researchers. The students met in groups until everyone had spent a full forty-five-minute session parsing both a dream and an emotional daytime event. Sharing a dream proved to be more helpful; scores on scales of exploration insight ("I learned more from the session about how past events influence my present behavior"; "I learned more about issues in my waking life from working with the dream/event"; "I learned things that I would not have thought of on my own") and personal insight ("I got ideas during the session for how to change some aspect(s) of myself or my life"; "I learned a new way of thinking about myself and my problems") were significantly higher if the students had worked with a dream.

Clara Hill, a psychologist at the University of Maryland, has studied how dream groups can help people improve a relationship or cope with a breakup. In one experiment, she and a coauthor recruited thirty-four women going through a divorce and invited twenty-two of them to a weekly dream group. Many of their dreams revolved around painful themes like failing or being thwarted or mocked. One woman dreamed of going home to reconcile with her husband and finding him in bed with two beautiful women in an apartment full of dead fish. Another woman dreamed of climbing a rope up a muddy hill, only to keep sliding back down. The twelve-person control group, meanwhile, spent the two-month period of the study on a waitlist before finally sharing their dreams in a single workshop. By the end of the experiment, the women who had participated in the ongoing dream group not only had gained insight into their dreams but also ranked higher on measures of overall self-esteem. The catharsis of sharing their secrets and the pleasure of belonging to a community translated into a confidence that stretched beyond the limits of the weekly dream group.

In another study, Hill and a colleague explored whether talking about dreams could help couples communicate about other issues too.

They recruited forty heterosexual couples — mostly college students — and assigned half of them to meet with a therapist for two dream-interpretation sessions. (The other twenty couples — the control group — were relegated to a waitlist.) Each partner would share a dream, and the counselor would guide the pair through a discussion of the feelings it evoked and how its meaning might reflect on the relationship. At the end of the experiment, the women — though not the men — in the treatment group felt that they had gained insight into their relationship and that their "relationship well-being" had improved. (Hill suggested that "the verbal sharing that was requested in the couples dream interpretation session may have appealed more to the women than the men.")

STUDIES LIKE THESE are useful in proving that psychologists should take dream groups seriously — but people don't need to consult the latest research to know that dream groups can be a source of insight and a balm for boredom and loneliness. Less formalized dream groups have cropped up as an organic bonding ritual in desperate situations. "Every morning we would start the day by sharing and interpreting the dreams we had during the night," one Auschwitz survivor wrote years after liberation. Dreams were a source of distraction in an environment sorely lacking in it; the dreaming mind was a self-reliant fount of entertainment. And the act of sharing dreams became an exercise in community-building. The Nazis replaced inmates' names with numbers and subjected them to barbaric conditions, but in sharing a dream or offering an interpretation, a prisoner could reassert his humanity.

"The interpersonal dimension of interpreting dreams in Auschwitz was connected with the inmates' need for capturing others' attention," Owczarski wrote. "When a prisoner shared an interesting dream, he or she became, at least for a while, important for his or her interlocutor . . . The meaning of a dream was not as important as the sheer fact of talking about it. Sharing dreams was therefore a kind of mutual help, aimed at increasing the inmates' self-esteem." In a vacuum of outside news, prisoners looked to dreams for clues to life-or-death questions like whether their relatives were still alive and whether the war would ever end. And because dreams were thought to contain prophecies pertinent not only to the dreamer but to other prisoners and the community at large, dissecting them was a legitimate group activity. Throughout the day, people could look for signs that an omen from another inmate's dream had been fulfilled. "When the dream did not come true for the dreamer, it came true for his friend," one prisoner said. "Dreams became common property: *see, your friend dreamt about it.*" They made up their own dream dictionary that reflected the precariousness of their lives and their preoccupation with the future. Smoking a cigarette prophesied the dreamer's release from prison. Cooking meat meant that he would be beaten during interrogation.

After liberation, many of the inmates were embarrassed to remember their one-time faith in dreams; the extreme stress of camp life had allowed them to suspend their disbelief. "It is hard to tell why we were all so naïve," one survivor wrote. "Nowadays, we see them [the dream interpretations] as immature or even silly, but back then they were simply necessary," said another.

For seven years, social worker Susan Hendricks led a dream group for inmates in a maximum-security women's prison in South Carolina. One day in 2005, the prison psychologist "went out and picked up two or three people on the yard and said, 'Would you all like to come and try this out?'" It took some work to convince the women to open up; in prison, "there's such a lack of trust, and fearfulness and suspicion." But as they got to know one another, they began to relax. When one woman was caught gossiping about the others' dreams, she was kicked out of the group; after that, the women felt even more confident that they could trust the remaining members.

The dream group helped satisfy their craving for community. "When you find a small group that you feel you can trust, that you can share things openly—there's a huge difference." Word spread, and by the time Hendricks left her position at the prison, dozens of women were on the waiting list to join. Within the confines of the dream group, they felt safe enough to become intimate in a way they rarely could. That trust transcended the boundaries of the formal meetings. "They'd see somebody from a distance, on the yard, and they might just wave or in some way acknowledge each other, and it gave them this feeling of connection," Hendricks said.

Many of the women struggled with brutal nightmares of violence or confinement. Nearly every night, one inmate would dream that she was wielding a shovel, digging aimlessly without knowing why. When she reached a certain layer in the soil, it would burst into flames, engulfing her. After bringing the dream to Hendricks's group, the woman realized that it had started shortly after her mother died. She had been granted leave to attend the funeral—but it had been a traumatic, humiliating experience. "They took her to the funeral in handcuffs and leg irons and the prison uniform. She was paraded in with guards on either side: they took her up to the front, had her look at her mother in the coffin, and then took her straight back to prison," Hendricks said. Once the woman recognized the connection between the nightmare and the funeral, it faded away.

Other members were able to use the dream group to prepare for stressful, high-stakes events. One woman came to the dream group the day before her parole hearing, distraught, desperate to talk about her nightmare: the hearing was a disaster; her request was denied. "The women encouraged her and told her their experiences going before the parole board and what she could expect. One of them said, 'I'll help you fix your hair.' Another said, 'We'll all be standing by to make sure you're okay.'" The next dream group was a week later, and the woman came with good news: she had won parole. By the following week, she was gone. "The dream-work smoothed out the terrible anxiety the dream had set in place, by giving her the support of her fellow group members who wanted to help her. Then she was able to be calm when she got in the room."

EVEN UNDER LESS extreme circumstances, dream groups can foster a much-needed sense of community and help people understand themselves more deeply. The Soul Dreamers of New York have been gathering for more than a decade, creating oases in a sometimes atomizing city. They have convened in apartments; they have rented studios in Chelsea; they have drawn funny looks at restaurants downtown. Their online Meetup group boasts over two hundred subscribers, though their ranks have ebbed and flowed over the years. Some members met at New Age retreats; others discovered the group online. A handful of regulars have held the club together. For them, the dream group is therapy, a hobby, and a social circle rolled into one. It's a refuge for people who take their dreams seriously in a world that doesn't understand.

On one of the first temperate evenings of spring, I met the Soul Dreamers at a dimly lit French bistro just off of Union Square. At the back of the restaurant, a crew of preppy twenty-somethings had gathered for some celebration; at another table, a well-dressed couple held hands while a teenage boy toyed with his phone. The hostess asked if she could lead me to the party of fifteen, but I shook my head and made my way toward a less glamorous group seated somewhat awkwardly —considering the intensity of the conversation already under way—in the middle of the cavernous dining room.

Michelle, a gregarious woman who seemed to be the group's leader, welcomed me to the table; she had been coming to these meetings for eight years. Another woman, a twenty-six-year-old student of psychoanalysis, said the friendly atmosphere at dream group was a relief —she and her classmates talked about their dreams all the time, but their analyses could get too competitive for her taste. Kristin, a painter and Reiki practitioner, had not owned a car since 2005, when she had a nightmare about crashing her father's old Saab; she was heading upstate to her family's home in the Berkshires when she realized her body was missing. "My consciousness was looking down on myself driving the car," she recounted matter-of-factly, as if she were telling me what she'd eaten for breakfast. "Then I saw the car on the shoulder, crashed and burned." She woke up, thought about it, and gave her car away. For years, Kristin had transcribed all of her dreams in the morning, and she had piles and piles of dream journals. "My whole life I've been a crazy dreamer. I'd go down to the breakfast table and want to tell everyone. They'd be like, 'Yeah, yeah.' I was that kid at the bus stop, telling people my dreams." She was in her sixties, and she'd realized that most people did not want to hear about her dreams. She was grateful to have found a community of like-minded people — people who preferred no other topic.

We ordered wine and cosmos, and Michelle explained the rules. Michelle had immediately been intrigued when she'd stumbled on the Soul Dreamers' webpage, but she'd had to steel herself to go to the bar where they were meeting. "Michelle, you have got to get over your shyness," she'd said to herself. "Just go." She became a regular, and she'd met some of her closest friends through the dream group.

The psychoanalyst-in-training shared her dream first: She was lying in bed, and a soccer star she admired was sitting with her, just hanging out. She felt disconcerted when she woke up: Why had the athlete been in her room, and why hadn't she realized that was bizarre? She gave the dream a title: "Out of Place." Irene had a kinder interpretation. Everything in a dream represented some aspect of the self, Irene reminded her. If it were her dream, she would think that she was connecting with a part of herself that she admired. The dreamer liked this interpretation.

EVEN PEOPLE WHO would balk at joining a regular dream group share their dreams in virtual communities, whether on designated dream apps or mainstream social media channels. When I see dreams pop up on my Twitter feed, they offer a moment of comic relief from the barrage of apocalyptic headlines. They're weirder, more spontaneous than most tweets — a reminder, amid the communal madness and hyperactivity of Twitter, that there are humans on the other end of the political pronouncements and self-promotion.

On apps like DreamSphere, Dreamboard, and Dreamwall, users log on to share their dreams with strangers and friends, "liking" and commenting on one another's dreams as though they were status updates on Facebook. An interest that can seem supremely individual, even solipsistic, becomes social, life-affirming — a reminder that even the oddness of dreams is universal. Something that made you laugh alone in the morning can make others laugh too. "I stabbed somebody with a bagel," confessed a user who identified herself as hhaalleeyy. "I died laughing," replied AwsamJournal — one of more than four hundred users who gave the dream a thumbs-up. "Funniest dream ever!!!!" wrote another. They share their glee, even their disbelief, over their first lucid dreams. "I had a body at times, other times not," reported Stratosynth in a post called "FIRST TRUE LUCID DREAM!!!" They crowd-sourced interpretations. "I had a dream my mum married a cockroach," one entry read. One amateur analyst wondered whether the dreamer felt betrayed by her mother or concerned about her mother's decision-making abilities. And they offer affirmation: Their dreams are not as strange as they think. "That's weird and kinda cute," one reader assured her. As lighthearted as many of these interpretations are, their volume is a sign of how intensely we want to understand our dreams.

NONE OF THESE stories or studies would surprise the members of IASD, many of whom have found their deepest sense of purpose and closest-knit communities by talking about their dreams. "The first time I came here, I was like, 'These are my people and I want to come here for the rest of my life,'" said Ange, the Canadian librarian. It's a sentiment I heard over and over. "These bonds are forever," said Victoria, an artist from Santa Fe. "Once you've shared a dream with someone, you can never undo that. It's not like you meet at a cocktail party."

During my own week at IASD, I found myself justifying the decision to wake up every day for the early-morning dream group. The day ahead would be hectic, and I wouldn't be able to write about the specifics without violating the others' privacy, but I set my alarm anyway. I got a glimpse into the psyches and preoccupations of strangers from all walks of life. I learned about a woman's complicated relationship with her late mother. I heard how a middle-aged nurse's self-image changed after she quit her job.

If I introduce the subject of dreams in my day-to-day life, whether at designated dream conferences or random parties, many people are eager to discuss them. In the course of writing this book, I have had countless fascinating conversations about dreams with editors and acquaintances, strangers and friends. "You must hear stories like this all the time," a new acquaintance will apologize, before launching into a dream that inevitably reveals some personal fear or fantasy and often leads to a deeper conversation; she discloses the addiction that once kept her from dreaming or the relationship she mourned in her dreams. My book topic has turned me into a magnet for confessions; I imagine I know how it feels to be a therapist at a party. I wake up to texts from friends about strange dreams they're dying to talk about. "I had a dream in which my boyfriend had a wedding ring on and I was like, WHY DIDN'T YOU TELL ME?" a friend messaged me the other day. I couldn't tell her what that meant, but we talked about how her last boyfriend had been going through a divorce while they were together. Another friend asked me why she'd had a dream about having sex with her dad. I had no idea, but we had a heart-to-heart about her anxieties about moving in with her (much older) boyfriend. These conversations bring us closer together, helping us to share stories and fears that might otherwise fester as secrets.

I've even persuaded some of my more skeptical friends that dreams contain useful psychological insights. When my friends agreed to trek to the Upper West Side to help analyze one of my dreams, they did it as a favor to help me with my book. But to everyone's surprise, the troop of us who learned about dream groups at Mark Blechner's office have continued to meet on our own for more than two years now. It's an unlikely gang, fluctuating as some members move away and others invite new friends. I had to cap it at fifteen, and we don't all hang out outside of our dream group. But one evening a month, like clockwork, we come together for a conversation more intimate than we usually have with our closest confidants.

It's less formal than the session with Mark; we drink cheap wine out of paper cups, order pizza, and sit on someone's floor. As we begin to digest, and after the ordinary conversation — the usual exchange of pleasantries, the catching-up, the *how-have-you-beens* — winds down, we follow the rules that Mark taught us. Whoever's turn it is passes out copies of her dream, which she's probably printed out at work and hastily collected from the communal office copier, and we go through the steps — reading the dream aloud, clarifying the manifest content, taking on the dream as our own, soliciting the dreamer's own associations and interpretation. Within this ritualized structure, we create a space where intimacy doesn't have to be earned, where disclosures flow as freely as bad wine. My friends say that making an event out of dreams has helped them remember their dreams more often, even changed their relationship with sleep.

When the group first started gathering, we tried to stick with safe dreams — funny, absurd, maybe opening up a conversation on something sensitive but more often staying in the comfortable realm of comedy. We brought in dreams that dwelled on concerns we were all likely to share, about our careers, about ambition and fear of failure. As the months wore on and we settled into a rhythm, we began to censor ourselves less. With a set framework in place, we could suspend the boundaries that usually govern conversation among acquaintances.

We gave one another tacit permission to probe, to ask the questions we wouldn't normally dare to. We've brought in dreams of sex and death and suicide; we've talked about childhood crushes and family secrets.

Even though she had only one friend in the group, S., a graphic designer, brought a dream to one of the first meetings she attended — a strange, convoluted dream that she couldn't stop thinking about. In the dream, she was walking through the cobblestoned streets of some foreign village, enjoying a lazy day out with an old friend, until her parents showed up and announced that they were going to drive her home. Her sister took over the wheel and dragged S. to a nail salon; annoyed, S. ran away from the spa. She came upon a polluted koi pond, saw an orange carp smoking a cigarette, fainted, and fell into the water.

"At first it was intimidating to share my dream," S. reflected later. "I didn't know everybody in the group well, and I knew I was sharing some personal details. Sometimes unpacking a dream can feel like airing your dirty laundry, but it felt nice that everybody responded genuinely and took the process seriously."

If it were her dream, Jo said, she would have enjoyed the first scene — walking through a new city — but would have felt stifled by the intrusion of her family. Someone else said she would be frustrated that she was always the passenger, never the driver. Only as the other women weighed in with their own associations did S. realize what, in retrospect, seemed obvious to her. "I lacked agency in this dream, and I often felt and behaved powerlessly. I was only really able to see that once it was pointed out to me." It was a relief to finally identify the feeling that dominated the dream, and she applied the insight she gleaned in dream group to the rest of her life. "I realized a lot of my familial relationships could be constrictive and bulldozing. Having and talking about this dream enforced that those feelings were valid and lasting. I became more conscious of how tied down I was feeling. From there I've been practicing saying (a guilt-free) 'no' more and also avoiding becoming a victim of circumstance."

"I dream a lot, and I pretty much always remember my dreams for at least a few days afterwards," said Jo. "I have nightmares, I have sleep paralysis. The dream world is something that's a big part of my life, something that feels significant to my emotional life, that I usually experience in a solitary way. So it feels both counterintuitive, but also extremely productive, to do something completely different."

"There's a formula and rules to it, which regulates all that intimacy but doesn't deaden it," said Moira. "It's so intimate. They're the kind of conversations you only have with your lover or your mom."

chapter 10

CONTROL

THE POSSIBILITIES I HAVE DESCRIBED SO FAR — rehearsing for real life, inspiring new ideas, balancing emotions, fostering community — can all be attained through regular dreaming. The evolutionary and cognitive functions of dreaming don't depend on lucidity.

But for people who learn to lucid dream, all of these benefits can be magnified. Those who master lucidity can dream about specific problems, seek answers or insight, stage cathartic encounters, and probe the recesses of the subconscious. Lucid dreamers have an edge in curing nightmares and in incubating memorable, emotionally rewarding dreams.

Not to mention the sheer pleasure of lucid dreaming. There is a remarkable consistency in the way people talk about their first lucid dreams, their induction into this unique state of consciousness. Efforts to describe it leave people grasping for words, resorting to clichés and nonsensical formulations; lucid dreams are "hyper-real," they're "more real than real." "The door handle is a door handle," as Steve Volk — the writer who suffered from recurring nightmares until learning to lucid dream — put it.

"Before having yourself tasted such delight, reader," Frederik van Eeden, the Dutch psychiatrist and lucid dreamer, teased more than a century ago, "you cannot imagine my elation when, on awakening, I found . . . that I had gone on observing—attentively observing, and thinking—thinking deeply and clearly, with full recollection and calm self-consciousness in that mysterious, senseless sphere of wonder and deception." Lucid dreamers are briefly liberated from their bodies, exempt from the normal laws of physics.

Ever since reading Stephen LaBerge's book in Peru, I'd had lucid dreams on occasion, but I couldn't predict when they would come; I got lazy about my reality tests, and I didn't always make time to meditate. Sleep was precious; waking myself up in the middle of the night was out of the question. Yet the more I learned about the power of lucid dreaming, the more I wanted to be able to induce lucid dreams on a consistent basis. I wanted to learn from LaBerge himself.

ON A HOT, humid day in September, I flew into Hawaii's tiny Hilo airport to find a bleary-eyed group already gathering. My fellow lucid-dream enthusiasts had picked one another out without too much trouble; they were the ones milling around sheepishly, looking a little rumpled, a little apprehensive, not quite sure what they had signed up for. I joined them and we waited for the shuttle, exhausting the browsing potential of the kitschy gift shop with its cheap leis and turquoise hoodies, swapping names and dreaming resumés.

"Have you ever had a lucid dream before?" a short, intense man asked me, wasting no time with pleasantries. His gingham shirt was mysteriously unwrinkled; he looked as though he might have gotten lost on his way to a business meeting. A yoga teacher was hunched over her phone, uploading pictures of palm trees foregrounded by the airport's parking lot to Instagram. A tall, wiry man with a faint Russian accent and a closely shaved head seemed to be in danger of falling asleep on his feet.

The conversation was running dry when a van bearing a sign for the

Kalani Oceanside Retreat Center pulled up. A chirpy young woman stepped out and shepherded us into the car. For the hour it took to drive to the resort, Natalie kept up a constant stream of chatter. I learned that she'd visited Kalani two years ago for what she thought would be a temporary vacation from her office job and had since made a new life for herself in Hawaii.

"People come here so introverted," Natalie gushed to the mostly quiet group, some of whom had just disembarked from twelve-hour flights. She promised us that by the end of our time here, we would have changed. We would be "so extroverted." We would be "different people."

Kalani is largely run by volunteers who live in tents and work half the week cooking, cleaning, and gardening for one another and for guests who come to learn yoga or ecstatic dance, ukulele or lucid dreaming. Some volunteer for a few months, passing through on gap years or personal journeys; others stay for years, alternating stints at Kalani with stretches earning more money at conventional jobs on the mainland.

The hour we spent with Natalie turned out to be a good transition, easing us into our full immersion in the physically uncanny, emotionally boundaryless world of Kalani. Wild boars roamed the property, and neon-green geckos clambered up the walls. Rugged volunteers puttered around in bandannas or straw hats, humming mantras to themselves or napping in hammocks, tossing off "I love you" and "You're beautiful" as casually as they said, "Good morning."

The whole district of Puna has a history as a magnet for seekers and searchers, a respite for pilgrims fleeing the pressures of modern life. Hippie co-ops and intentional communities are dotted across the area. So-called Punatics wander the black-sand beaches in dreadlocks and ratty clothes and loiter in the hot springs, smoking.

Natalie showed me to my room, a simple, dormitory-like space with hokey pastoral paintings on the walls, a few pieces of wicker furniture, and little else. The primary source of light was a single bare bulb on the ceiling, but the electricity was out that night. I stumbled around with the tiny flashlight on my key ring and passed out.

When I drew back the flimsy curtains in the morning, I took in the scene properly for the first time. From my window, I could see luscious palm trees and tall tropical grasses misted over by a layer of fresh dew. My first thought was that the landscape resembled a desktop background come to life.

I DON'T THINK I have ever been part of a group so diverse on nearly every axis I can think of. Some had traveled from as far away as London and Australia, some from as close as the West Coast and Hawaii itself. One man saved his pension and skimped on his biggest expense—marijuana—to scrape together the tuition money; others' trips were sponsored by their parents. There were twenty-something women who looked like they had just stepped off a resort-wear catwalk, with seemingly endless wardrobes of sundresses and sarongs for day, wraps and shawls for evening. There were old men who seemed to possess only tropical-print shirts. Some had barely heard of Stephen LaBerge; others had worshipped him for years, had practically memorized his books and techniques for lucid-dream induction. Some had never experienced a lucid dream; a couple had been lucid dreamers all their lives. If they had anything in common, it was a fluency in the literature on near-death experiences and extrasensory perception.

For Jules, who quit her job in TV production to travel the world teaching yoga, dreams fit into her journey ever deeper into her own psyche. Alana, an accomplished meditator and veteran of sensory-deprivation tanks, had recently graduated from NYU with a major in millennial storytelling—a program she designed herself to learn about "how millennials tell stories differently"—and hoped to use lucid dreaming to locate the wise woman within her. This was her second pilgrimage to Kalani, and she relished her role as an unofficial guide to the resort. *Where is the meditation tent? Can you wear a bathing suit at the nudist pool?* (She didn't recommend it.) Seventy-year-old

Michael hoped that if he learned to recognize when he was dreaming, he would stand a better chance of noticing when he arrived in the afterlife. It was a morbid project, but he had a sense of humor about it. "Ticktock," he said. "I'm cramming for finals."

IN THE MORNING, we convened in a bright, airy structure on top of a hill, one side opening directly into the rainforest. Knotted scarves hung from the window frames, and a portrait of the volcano goddess Pele, painted in fiery primary colors, dominated one of the eight walls. I have never found four-walled rooms particularly stifling, but this space had been designed, according to Kalani's promotional literature, to liberate visitors from "box-based architecture."

Stephen's assistant, Kristen — a clinical psychologist and master lucid dreamer with the pun-embossed T-shirts and upbeat mien of a camp counselor — regaled us with tales of her lucid adventures. Kristen taught herself to induce lucid dreams in college after she learned about the phenomenon in a psychology class. "I couldn't believe it wasn't a commonly known thing," she said. "I was just so in awe." She had since trained herself to become lucid as often as three times a week and could even meditate and practice yoga in the dream state. She was outlining our curriculum for the week when she was interrupted by a low-pitched masculine shout.

"What are we doing here?" bellowed a barefoot man in a baggy Hawaiian shirt and shorts, bright blue eyes peering out from beneath bushy white eyebrows. Stephen must have slipped in through the back door while Kristen was talking; I had missed his entrance. His voice swung theatrically; each question began as a rumble and ended as a squeal.

"What is this all about?" he demanded. "How do I know you're people? Maybe you're robots or aliens or dream figures. Does anybody think that it really might be a dream?"

This barrage of questions was a fitting introduction; Stephen would spend much of the coming week training us to pay closer attention to our surroundings, to scrutinize the details of our environment, to search for incongruities and stop assuming that we were awake. He greeted us one by one, mustering an impressive show of curiosity over each person's individual path to Kalani. At sixty-nine, he had devoted the better part of his life to lucid dreams, and it was "re-vivifying," he said, "to be with people who find the topic intriguing."

Stephen was intense in a way that a sympathetic observer might describe as cerebral; a less generous one might have characterized him as awkward, even manic. He was constantly in motion even when he was sitting, contorting his body this way and that, crossing and uncrossing his ankles. When he got excited — which was often — he jumped out of his chair. His gesticulations sometimes devolved into jazz hands, and his voice could cover several octaves in a single sentence. More than once, I heard his manner likened to that of a wizard.

Lucid dreaming has been slowly gaining prominence in recent years. The release of Christopher Nolan's 2010 science fiction blockbuster *Inception* — in which corporate spies sneak into their marks' dreams to steal their secrets and implant bad ideas — was a landmark moment. (The spies use a top as a tool for reality tests; if it spins indefinitely, then they know they are in the dream state; if it falls, they are awake.) Nolan said that the film was inspired by his own experience of lucid dreaming and that its ambiguous ending — the camera lingers on a spinning top, leaving viewers to wonder whether or not it will fall — should be taken to mean that "perhaps all levels of reality are valid." Google searches for *lucid dreaming* spiked around the movie's release and have never returned to pre-2010 levels. And the internet, of course, has helped. A constantly updated Lucid Dreaming forum on Reddit has accumulated more than 190,000 subscribers.

Still, lucid dreaming has not exactly permeated the culture. And the lucid-dream buffs at Kalani could hardly be called adherents of the mainstream.

Locks of chin-length gray hair hung in Michael's face, and he stared, unblinking, through rimless glasses. He had a habit of cocking his head to one side and leaving it there for a while, as though he'd forgotten where he put it. He was not, after all, accustomed to this much stimulation. His speech was slow, his cadence monotonous, as he explained how he wound up living in a state of near seclusion. He worked for thirty years as a psychiatric technician on the West Coast; he and his late wife left for Mexico on a whim. "She said, 'Let's take an early retirement and have an adventure,'" he recalled. "It would never occur to me, but I'm ready to jump into the unknown." Mexico had proven a suitable location for Michael's second act. "I live in the jungle in an isolated house—a perfect area for me not to be bothered by neighbors," he said. He spent up to three hours a day in silent meditation.

Michael had been fascinated by dreams ever since he was a teenager, when he decided he was neurotic and set out to educate himself about psychology. He had short-lived infatuations with Freud and Jung, but it was Asian philosophy—and Asian-inspired superstition—that stuck. For forty-five years, Michael had been running his life according to the decrees of the I Ching, an ancient Chinese system of divination that generates meaning-laden hexagrams based on coin tosses. He asked the I Ching whether or not he should splurge on the trip to Hawaii, and it showed him the symbol for a well, so he ran to his computer and purchased a plane ticket. The main constant in Michael's life—more than any single person, more even than Eastern spirituality—had been pot. He'd been smoking for half a century, sometimes up to four or five joints a day. He was often stoned at work but believed it helped him empathize with psychotic patients. "I did some of my best work while I was high," he said proudly.

All these years later, though, he didn't want to be dependent on a mind-altering plant anymore. "Now that I have no responsibilities, my chief worry is that I'm still smoking marijuana," he said. He had tried to quit a few times, but sobriety had never stuck. He hoped that if he could learn to lucid dream, he could confront whatever demons were shackling him to his vice. It would also be nice, he figured, to "realize the illusory nature of my worries and concerns in ongoing existence."

On the first day, I pegged Pavel for the strong, silent type. By lunch on the second day, it was clear that my initial impression was wrong; he must have just been sleepy. Pavel came alive when he talked about the self-help tricks that had changed his life. He abstained from meat, dairy, and alcohol and often wore a five-hundred-dollar headset that delivered mild electric shocks to his scalp. He idolized a man who'd trained himself to replace full nights of sleep with periodic twenty-minute naps. If I would only learn to juggle, he assured me, I would write my book five times faster. With every new life hack he described —holotropic breathwork, sensory-deprivation tanks, silent meditation retreats—he grew ever more animated; he made me think of a wind-up toy or a top, spinning faster and faster, and I wondered what would happen when he stopped—would he turn into a cyborg? A fully optimized machine? What was he programming himself for?

Even the participants with only a loose connection to what most of us call reality were aware—mostly—that their hobby was not a topic for general conversation. Theresa, a fashion photographer who self-identified as "high-energy," told only a few close friends that she was coming to Kalani. "When I tell people I'm really into lucid dreaming, I feel judgment of, like, 'You're crazy,'" she said. "To most people, I just said I was going on vacation in Hawaii."

A neurologist and an entrepreneur from Boston signed up for the retreat to learn more about the market for a lucid-dream induction device they wanted to build. After three days, they changed their flights and left. That was around the time I began flouting Kalani's rules and using the limited bandwidth to stream NPR's *Morning Edition* on my laptop, never mind the paper-thin walls. I needed just a few moments of normalcy.

BY THE END of his painstaking period of trial and error as a student at Stanford, Stephen had created a powerful system that not only enabled him to lucid dream whenever he wanted but worked for other people too. The core of Stephen's method, the sine qua non, is what he calls the reality test. Aspiring lucid dreamers should make a habit of asking ourselves at regular intervals throughout the day whether we are awake or asleep. Because daytime routines work their way into dreams, we should pose the same question in our sleep. If we are sufficiently attuned, we will respond that we are asleep, and a lucid dream will commence. (Back in the 1960s, the mysterious New Age author and anthropologist Carlos Castaneda related a similar technique, which he claimed to have learned from a Mexican shaman: Study your hands during the day and ask yourself if you are dreaming; when you notice your hands in a dream, you should realize that you are asleep.)

Effective reality tests entail reorienting yourself in the world, cultivating a skeptical outlook toward your environment. Is everything as it should be? Look for clues that your surroundings might not be real. Inspect your hands — does each one have the usual number of fingers? Check the clock, and check it again — has a reasonable amount of time elapsed? Find a shiny surface — are you reflected back as you really are, or are you distorted, as though you're looking in a funhouse mirror? Jump up in the air — do you drop back to the ground, or have you suddenly acquired the ability to fly? The dream world is constantly in flux; check whether your environment is stable. Exit a scene and then return to it. Are you in a different room? Find a piece of text — the spine of a book, a word on a bracelet, an e-mail — look away from it, and then look back. If you're in a dream, the words are likely to have changed by the second inspection.

"Does anybody think that this might be a dream?" Stephen demonstrated a reality test.

Silence; we glanced sideways at one another, like students taken aback by a pop quiz.

"Are you sure you're not going to wake up in bed in another ten minutes or an hour?"

Tentative nods of assent.

"But how do you *know*?" Stephen asked. "What is the evidence for that assumption?"

"I can't float," one brave guy called out. He was sitting motionless in his chair.

"You call that trying?" Stephen shouted. His incredulity was melodramatic, his voice rising in a show of outrage. "That's not a real effort!" Stephen straightened his back as though trying to levitate out of his chair, his face crumpling with the strain of his imagined effort. He jumped up, his eyes widening as if in hope. But he dropped back down into his seat; he could not float. He was awake, and he had conveyed his point. A proper reality test entails truly considering, with your body as well as your mind, the possibility that you are in a dream.

Reality tests are as idiosyncratic as dreams; what works for one person may not work for another. The reality test that involves flying is best for people whose dreams are already active. The one that relies on reading is appropriate for more literary dreamers. Another factor is the dreamer's tolerance for disrupting her daily life and drawing attention to her unusual mission. If you don't mind looking slightly deranged, you can try to push the finger of one hand through the palm of the other (if your hand is permeable, you are asleep) or pinch your nose and try to inhale (if you can breathe through your plugged nostrils, you are dreaming).

Most beginners aim to perform a reality test about ten or twelve times a day. Some find it useful to set hourly alarms; others prefer natural prompts, examining their state of consciousness every time they perform certain routine actions, like walking through a doorway or looking in a mirror. Stephen gave each of us a physical mnemonic device — a blue wristband printed with the mandate AWAKEN IN YOUR DREAM — to use for our reality tests. For the rest of the week, people zoned out midconversation to ogle their own wrists, lift their eyes to stare into the middle distance, and then look back at their arms, checking whether the bracelet still bore the same message.

Learning to lucid dream can infiltrate every part of the day, from dream journaling in the morning to questioning the state of reality throughout the day, meditating before bed and waking up at strategic points in the night. The more time you spend contemplating dreams, the more you will unite your sleeping and waking worlds, bringing dreams into the daytime just as you hope to bring conscious thought into dreams. Stephen encouraged us to transcribe our dreams in as much detail as possible whenever we woke up, to cultivate dream recall — a prerequisite for lucid dreaming (he suggested we remember at least one dream per night before attempting to lucid dream). This exercise should generate a robust sample of our own dreams; by analyzing our dream diaries, we should be able to pick out our personal dream signs — recurring motifs or features that help us recognize when we're in dreams. It could be an anomaly in the environment or in how you perceive it. For some people, the giveaway is a certain distorted object or far-fetched scenario; for others, it's the arrival of a long-dead relative or extraterrestrial creature. One of Stephen's dream signs is that a contact lens has fallen out of his eye and started "multiplying like some sort of super-protozoan." One of mine is the realization that I'm a passenger in a driverless car; another is that I can't stop unpacking a suitcase that perpetually refills itself. Sometimes, though, I become lucid when I notice an unlikely situation that's harder to categorize: Why won't these twenty people in my bed stop talking and let me sleep? Why have all the files disappeared from my computer?

Just as intention helps to improve dream recall, it's also an important factor in becoming lucid. Stephen urged us to think about lucid dreams as often as possible, to discuss our dreams over meals and between lectures. When our afternoon lessons ended, he showed us dreamy movies like *The Truman Show* and *The Last Wave*. In the morning, we broke into groups to recount whatever dreams we could remember, to reflect on dream signs we had missed, and to congratulate those with lucid dreams to report. The events of the day were so odd that our reality tests were almost reasonable; we had created an environment where the typically rigid boundaries between asleep and awake were looser and where it made sense to wonder whether we might be dreaming.

In the evenings, we meditated — a practice correlated with lucid dreaming as well as overall dream recall. Meditation is like cross-training the brain for lucid dreaming, honing general mental agility and awareness in the hope that the mindset carries over to the dream state. In 1978, psychologist Henry Reed found that regular meditators had clearer memories of their dreams on the days after they had meditated. More recently, psychologist Jayne Gackenbach also noticed a difference in dream recall among meditators and nonmeditators; among the 162 college students in her study, those who regularly meditated could remember, on average, 6.2 dreams per week while those who didn't recalled 5.1 dreams over the same period.

It's easiest to become lucid late in the night, when REM phases are longer and dreams are already more story-like and intense. One of the most reliable induction methods targets those final sleep stages. Stephen taught us to set an alarm to wake ourselves up after either three or four REM periods — about four and a half or six hours after going to sleep. (The super-attuned could set an intention instead of an alarm, waking themselves at the desired moment through sheer force of will.) Stay awake for thirty to sixty minutes: Get out of bed and do some quiet activity, like reading, preferably books about lucid dreaming. Write down the dream you just woke from and replay it in your mind; relive it over and over, until you know it by heart. Then, imagine that you are dreaming it again, but this time concentrate on a moment when you could have become lucid, a dream sign you missed — if only you had noticed that you had wings, say, or that your friend was the size of a thimble — and then let yourself fall back to sleep. These later, longer REM periods are inherently more fertile territory for lucid dreams, and, as Stephen pointed out, you'll maximize the odds of remembering your goal to become lucid. "It's easier to remember to do something two minutes from now than two hours from now."

In another induction method, the dreamer transitions directly from a state of wakefulness into a lucid dream. If you home in on the hallucinatory hypnagogic images that accompany the transition from wake to sleep, you might follow the image until it turns into a dream. "If you keep the mind sufficiently active while the tendency to enter REM sleep is strong, you feel your body fall asleep, but *you*, that is to say, your consciousness, remains awake," Stephen wrote. "The next thing you know, you will find yourself in the dream world, fully lucid." These wake-induced lucid dreams are usually possible only toward the later stages of REM or during daytime naps, when the dreamer is primed to slip quickly into REM sleep. "Try to observe the images as delicately as possible, allowing them to be passively reflected in your mind as they unfold," Stephen advised. "While doing this, try to take the perspective of a detached observer as much as possible." Static images should come together in sustained sequences, and as they become more vivid, "you should allow yourself to be passively drawn into the dream world." This technique has never worked for me — I find hypnagogic images relaxing, and I just fall asleep — but some people who are more skilled than I am at meditation swear by it.

Inexperienced lucid dreamers are often so excited to find themselves in a dream that they wake up. But once you have entered the lucid sphere, there are techniques that can prolong the dream. When Stephen was first experimenting with lucid dreaming back at Stanford, he would sometimes slip into a regular dream or just wake up upon becoming lucid. But if he could create the feeling of being physically present in the dream, he discovered, his odds of remaining in the dream state improved: "Just being engaged is enough to stabilize it." To locate his dream-body, he experimented with various tricks, like rubbing his hands together and twirling around. If he splayed his arms and spun like a top, he could usually orient himself in the dream world long enough to carry out his goals. Another method he found useful was to repeat a mantra within the dream, like "This is a dream, this is a dream" or "I'm dreaming."

It's a misconception that anything is possible in the lucid-dream world; it's easier to influence the dream than to fully control it. Rules in the dream world are more flexible than in the real world, but they still apply, and they vary from person to person. One lucid dreamer might be able to skate but not fly; one might be able to change the weather but not the environment. Not all obstacles can be pushed aside, and, as in life, temptations like sex can distract the dreamer from her goals. One of the most confounding qualities of lucid dreams is that other characters seem to have independent agency; also as in life, it's rarely possible to control the behavior of other people.

THE PREVALENCE OF lucid dreaming is difficult to measure, and inconsistencies in definition complicate the issue; lucidity is a spectrum, and scientists disagree about the level of consciousness and control implied by the term *lucid.* Must a lucid dreamer have full command of the details of her waking life? sticklers ask. Must she be able to control how the dream unfolds — or is an inkling of awareness enough? ask the more lenient.

Regular lucid dreaming appears to be most common in children and adolescents; one study suggested that six- and seven-year-olds had lucid dreams much more often than older children and adults. Other research indicates that people who lucid dream easily have an unusually high "need for cognition" and a strong "internal locus of control"; they tend to think things over and believe that they are responsible for what happens to them. A handful of studies that are very popular on lucid-dreaming forums suggest a link between lucidity and creativity.

Lifestyle choices and hobbies can also have an effect on dreaming patterns. Video-game enthusiasts tend to have more lucid dreams than nongamers; even their non-lucid dreams are more outlandish, incorporating supernatural or extraterrestrial scenes. In what may be a self-perpetuating cycle, both gamers and lucid dreamers have better-than-average spatial awareness and a lower susceptibility to motion sickness. "The major parallel between gaming and dreaming is that, in both instances, you're in an alternate reality, whether a biological construct or a technological one," Jayne Gackenbach said. The same principle might explain why athletes enjoy a high lucid-dreaming rate — the time they spend mentally practicing their sport could prepare them to take control in their dreams. (One German study involving hundreds of professional athletes found that about 14.5 percent of the dreams they remembered were lucid, compared to 7.5 percent for the general population. What's even more suggestive is that the majority of lucid-dreamer athletes — 79 percent — began lucid dreaming spontaneously, without making any special effort.)

The most comprehensive review so far — a 2016 meta-analysis incorporating fifty years' worth of studies and more than twenty-four thousand respondents — concluded that 55 percent of people have experienced lucidity at some point, and nearly one-quarter have at least one lucid dream per month. Scientists who work with lucid dreamers tend to make similar estimates. People who can become lucid any night they choose — a feat requiring a terrific degree of control — are scarce enough that researchers are always hunting for them. Dutch cognitive neuroscientist Martin Dresler, who has spent about as much time recruiting, interviewing, and observing lucid dreamers as anyone alive, offered a few impressions of that rare dreamer who can become lucid at will. "Most of our subjects have quite some self-discipline," he said. "Most of them don't drink coffee, don't drink alcohol, don't smoke."

Stephen — who happily admitted, beer in hand, that he did not conform to Dresler's definition of a disciplined person — maintained that individual differences like gender, age, personality, and diet pale in comparison to the single trait shared by all lucid dreamers: high dream recall. Adolescents may have more frequent lucid dreams, but they're also more likely to remember their regular dreams.

Stephen's books strike an encouraging tone; "You can do it!" is the takeaway, if not the guarantee. But as with other activities that demand subtle mental mastery, like meditation and mindfulness, those who fail can be faulted for trying too hard. Dresler told me about one research subject who trained for six months to have a lucid dream but couldn't manage it; only after he stopped trying did he succeed.

For all of Stephen's optimistic inclusivity, the question of whether everyone is capable of lucid dreaming has not been settled. Lucid dreaming doesn't come naturally to everyone. I have met several people — some whose fruitless struggles had taken them all the way to Kalani, others whose frustration only motivated them to corner me at parties and ask for tips — who, in spite of extravagant effort, have never succeeded, people who insist they have diligently implemented all of the techniques Stephen laid out, to the letter or beyond, performing hourly reality checks, even at the expense of social graces; reconfiguring their sleep schedules; meditating in the middle of the night. And the regimen we followed at Kalani didn't work for everyone. Three or four people had their first lucid dreams on the retreat; I had more lucid dreams during the workshop than I've ever had in a single week. But a few of the people who seemed the most desperate to lucid dream never did.

In 2012, a team of European psychologists led by Heidelberg University's Tadas Stumbrys sifted through the academic literature on lucid-dream induction and found thirty-five papers dating back to 1978. Some involved specific populations, like children and nightmare sufferers; some relied on the classic groups of undergraduates. The studies included induction methods that are now the go-tos — reality testing, meditating on the desire to become lucid — as well as more far-fetched ones. In 1978, British psychologist Keith Hearne tried to induce lucidity by spraying water onto sleeping college students' faces with a syringe. A few years later, he administered electric shocks to his sleeping subjects' wrists.

On the whole, the team found the field to be plagued by poor methodology. Sample sizes ranged from ninety-four subjects to only four. Nor did Stumbrys find any of the induction techniques to be entirely reliable. After analyzing the data, he concluded that "most lucid dream induction methods produced only slight effects, although some of the techniques look promising." He found the best evidence for cognitive techniques — like setting an intention and reality testing — rather than applying an external stimulus or taking a supplement or drug.

In most of those studies, though, results were tracked only over a few nights; subjects did not have time to practice the kind of subtle methods Stephen prescribes. In 2016, David Saunders, a lecturer in psychology at the University of Northampton in England, designed a longer, more naturalistic study of lucid-dream induction. He recruited people who had experienced no more than one lucid dream in the previous three years and assigned fifteen to a control group and twenty to an experimental group. The experimental group received coaching in Stephen's techniques; they were taught to keep a dream diary, search their journal for dream signs at the end of each week, practice reality testing, and meditate on their intention to become lucid. Like the group at Kalani, they were given wristbands inscribed with the question AM I DREAMING? and instructed to use them for reality tests throughout the day. Saunders checked in with his subjects over the phone every week, reminding them of their assignments and asking about their dreams. Over the course of the twelve-week study, nine members of the experimental group — 45 percent — succeeded in inducing a lucid dream. The bracelet turned out to be the most effective aspect of the program; six of the nine people who achieved lucidity were triggered by looking at the wristband in their dreams.

New research continues to provide encouraging signs. In an Australian study published in 2017, 169 adults practiced different induction techniques, and within just one week, 45 percent of them had had at least one lucid dream. The most effective strategy was to combine regular reality testing with waking after five hours and repeating the goal of becoming lucid while falling back to sleep; 53 percent who followed those instructions succeeded in having a lucid dream.

The relationship between effort and success is not linear; many people begin lucid dreaming without even meaning to. You might have a lucid dream as soon as you learn that it's possible; you might implement Stephen's exercises for weeks before catching a glimmer of awareness; you might practice in vain and succeed only after you've given up. "We have some hints that it might be possible for everyone," said Dresler, but for now, "we don't have the data."

STEPHEN DIDN'T START leading retreats just to pay the bills or even to share the joys of lucid dreaming. The workshops have also provided him with a way to move his own research ahead. They have given him access to a group of people who are willing to participate in his studies, even if they aren't certified by a lab.

This year at Kalani, that tradition continued. On three consecutive nights, those of us who agreed to take part in what Stephen cryptically called "the experiment" were given plastic bags of unmarked, oversize capsules and instructions to swallow them after our third REM period, meditate or write in our dream journals for thirty to sixty minutes, and go back to sleep. The three packets contained one set of placebo pills and two of galantamine, a drug developed to treat Alzheimer's disease. (It's available both over-the-counter and as an FDA-regulated prescription.) Alzheimer's patients suffer from low levels of neurons that respond to acetylcholine, a chemical that sends signals between nerve cells; the imbalance can contribute to their lapses in memory. Galantamine — one of a number of drugs classified as cholinesterase inhibitors — works by preventing the breakdown of acetylcholine in the brain. Bizarre dreams are a side effect; galantamine reduces "REM sleep latency," the time between sleep onset and the first REM stage, and increases "REM density," a measure of eye-movement frequency that corresponds to dream intensity.

Galantamine should enhance mental clarity in the dream state in the same way that it improves memory in dementia patients. Over the years, Stephen has served different doses of galantamine and other cholinesterase inhibitors to over one hundred aspiring lucid dreamers. His results are promising; he has found that people who are already proficient lucid dreamers are five times more likely to become lucid on the nights they take galantamine than on the nights they take a placebo. Even without yet publishing these findings in a peer-reviewed journal, he has — thanks to presentations at IASD and word of mouth — helped set off a wave of formal and informal research, stimulating the market for lucid-dreaming supplements with names like Galantamind. Online lucid-dreaming boards are teeming with inspirational stories of galantamine-assisted success. "The first night I took it I had one lucid dream after another," wrote a member of the World of Lucid Dreaming Forum. "Most of the times I take it, I have outstanding dreams — often flying dreams and amazing journeys that blow my mind," another attested. One researcher surveyed nineteen lucid dreamers who incorporated galantamine into their routines and found qualitative differences in the way they described their drug-fueled lucid dreams: they were more vivid, longer, and more stable than usual.

Galantamine is not a magic bullet, though; it can trigger nasty side effects, like headaches, nausea, and insomnia. And it can work too well — cautionary tales of galantamine-induced nightmares can be found alongside success stories. "It felt like my brain was being drawn and quartered," one lucid dreamer wrote. "I kept falling back asleep into these bizarre dreams that I can only describe as my head being scraped against the bottom of a submerged iceberg." "It felt like I was falling through my bed and all these loud screeching sounds and vibrations started happening," testified another. "It was so scary and I felt paralyzed."

The day after our experiment began, a few people turned up to the morning lecture looking haggard and complaining that they hadn't been able to fall back to sleep after taking their pills; one had spent the night vomiting. For me, galantamine did the trick. On both of the nights that I took it, I had lucid dreams, and no trouble falling back to sleep. When I took what I later found out was a placebo, I could recall only a mundane, non-lucid anxiety dream in which I found out that an acquaintance was also working on a book about the science of dreams. What I think was more helpful than galantamine, though, was the fact of being on the retreat — in a place where I didn't have to think about everyday things and where I was surrounded by people who shared my goals. I don't think it was a coincidence that my first lucid dreams in Peru came at another time when I was able to maintain a single-minded focus on my desire to become lucid and when I had made dream talk a regular part of my day.

WHILE STEPHEN HAS been doggedly plowing ahead on his own — assembling groups of ten or twenty in exotic locales, tallying his results back home in Arizona — the field he helped launch has been making real strides, finally becoming respectable within the sciences. EEG studies have revealed that areas of the brain that are dormant during regular sleep are switched on in lucid dreams. In 2009, German psychologist Ursula Voss found that the frontal lobes — which are involved in higher-order cognitive processes like logical reasoning, problem-solving, and self-reflection and which typically shut down in REM — light up during lucid dreams. She designated lucid dreaming a "hybrid state of consciousness," combining elements of waking and sleeping cognition.

Efforts to replicate and elaborate on Stephen's early works have confirmed some of his findings but complicated others. Attempts to replicate Dement's discovery of the link between dream actions and eye movements — the discovery on which Stephen's proof depended — yielded mixed results, and the so-called scanning hypothesis became a subject of controversy. Critics pointed out that babies' eyes flutter during REM, even though they are incapable of visual dreaming, while proponents held up studies of the blind. People who lose their sight before the age of around five can't see in their dreams; those who become blind later in life retain some capacity for visual imagery, though the pictures fade from their dreams over time. In the 1960s, psychologists in Chicago studied the sleep of jazz pianist George

Shearing, who had been blind since birth, and noticed that his eyes barely moved during REM. More recently, Peretz Lavie, who founded the sleep lab at the Technion–Israel Institute of Technology, studied people who had lost their sight at different ages and discovered that the longer his subjects had been blind and the less sight they consequently had in their dreams, the less their eyes moved during REM.

Daniel Erlacher and Michael Schredl have also shown — like Stephen — that counting takes about the same amount of time in lucid dreams and in waking life, but their team has noticed that physical tasks — like walking, doing squats, or performing a brief gymnastic routine — take slightly more time in the dream state. "Longer durations in lucid dreams might be related to the lack of muscular feedback or slower neural processing during REM sleep," they suggest. Stephen's studies of hand-clenching and hemispheric lateralization haven't been replicated, but there are hints that they could be. In one experiment, Martin Dresler and his colleagues asked six expert lucid dreamers to squeeze their fists in a dream. Only two managed to carry out this assignment, but for both of them, the dream-clench did correlate with activation in the sensorimotor cortex — the same regions involved in real-life fist-making. Dresler did not, though, measure the muscular movement of their actual hands. "Previous studies have shown that muscle atonia prevents the overt execution of dreamed hand movements, which are visible as minor muscle twitches at most," he wrote. One small study, inspired by Stephen's work, explored the possibility of translating dream–hand clenches into real ones. As part of his doctoral dissertation, Erlacher asked lucid dreamers to try to open and close their hands in the dream state while an EMG measured muscle activity in their forearm. "We found in some participants a small EMG activity, as in Stephen's original paper," he told me, "but sometimes we did not find any EMG activity." Even so, Erlacher said, "Stephen's early studies are very important. He's a very inspiring person."

• • •

SCIENTISTS ARE FINDING powerful applications of lucid dreaming for intellectual as well as therapeutic and clinical problems. "If you want to study subjective experiences and their neural correlates, dreams are an excellent means to do that," said Katja Valli, a neuroscientist at the University of Turku in Finland. She believes that pinpointing the neural differences among dreamless sleep, dreams, and lucid dreams could shed light on the cognitive basis of consciousness itself. Dresler hopes that learning to become conscious and take control in the dream state could help schizophrenics recognize the hallucinatory nature of their psychosis. "In normal dreaming and schizophrenia, people lack insight into their current state," Dresler said. During non-lucid dreams, even the most mentally stable people draw random connections and lose their sense of perspective; the dreaming brain bears some resemblance to the psychotic brain. But in lucid dreams — when people are conscious but may lack full volition — he sees "a model for impaired insight." If schizophrenic patients could master lucid dreaming, then during their next acute phase, "They might have a better chance of realizing they are in an impaired state."

Lucid dreaming can also help people with more common mental disorders, like anxiety. German psychologist Paul Tholey intuited the therapeutic potential of lucid dreaming back in the 1960s. After the death of his father — with whom he had had a complicated relationship — Tholey was haunted by father-like figures who bullied and berated him in his dreams. When the paternal demon showed up in his lucid dreams, he reflexively took the opportunity to attack, sometimes succeeding in transforming his father into a less imposing creature — often a dwarf, occasionally a mummy. But the satisfaction he derived from these victories always faded by morning, and the menacing figure kept coming back. One night, Tholey decided to try a different tactic. The figure began his usual approach, but instead of lashing out, Tholey engaged him in conversation. He reprimanded his father for stalking his dreams but conceded that some of his criticisms were fair, and the two shook hands. "This lucid dream had a liberating and encouraging effect on my future dreaming and waking life," Tholey wrote. "My father never appeared again as a threatening dream figure."

Hoping to extract a more general principle from this experience, Tholey began running experiments on his students. He suspected that people could improve their mental health by deliberately conjuring up problems and confronting them in the safety of the dream state. In one set of studies, he instructed lucid dreamers to seek out a frightening person or situation within the dream. If the dreamer found herself floating on top of a pond, she should deliberately drop to the bottom. If she wound up in an open meadow, she should look for a dark forest. Once the dreamer had located an adversary, she should resist the temptation to attack and attempt to make peace instead. After setting this task to sixty-two lucid dreamers, Tholey managed to collect 282 dreams that featured a menacing character; in one-third of them, the dreamer succeeded in reconciling with the figure. One effective technique was the one he had discovered himself — engaging the foe in friendly dialogue. Looking directly into the adversary's eyes could also lead to a truce. And as Tholey had hoped, the afterglow of these imaginary détentes lasted well beyond the dream — 62 percent of Tholey's subjects said they felt less anxious in their waking lives, and 45 percent testified to feeling more emotionally balanced.

Line Salvesen has been both an anxious person and an effortless lucid dreamer for almost as long as she can remember. As a child, she suffered from recurring nightmares and realized that she could escape from them if she recognized that she was in a dream. In one, she would be riding in the back seat of a car when all of a sudden, her parents — who were driving — would vanish. The car would hurtle down the road, toddler Line powerless in the back, until it crashed. She figured out that she could wake herself up, which helped, but it was only after she taught herself to seize control that she was able to banish the nightmare for good. One night, after her parents disappeared as usual,

Line consciously formulated a new plan: She would summon her kindergarten classmates to steer the car. "They were in the driver's seat, and they helped each other," she said. "It wasn't really a nightmare anymore."

It wasn't until reading an article about lucid dreaming in a magazine that Line — who was having lucid dreams almost every night — realized that not everyone was conscious in dreams. "It said that only a small fraction of people are able to have these naturally, and I was like, 'I'm special?'" She laughed. The habit that was as intuitive for her as breathing, she learned, was an elusive goal for others.

In spite of her special skill, Line suffered from overwhelming anxiety in her teens and early twenties. "I felt stressed all the time," she told me. "I didn't feel that I had any control." She tried therapy and medication, but nothing worked. "It made life pretty hard," she said. "It ruined my senior year in high school." She missed classes because she was so tired — even though she was sleeping twelve hours a night — and her grades plummeted. She took sick leave from her first job to undergo more intensive treatment.

Until Line met lucid dreaming expert Robert Waggoner in a cyber-dreaming conference, she had mostly used her lucid dreams for fun, but Waggoner suggested they might hold the key to solving her anxiety. The next time she became lucid, she followed his advice. "I told myself that I would be happy and anxiety-free for one week. I just said it out loud in the dream, with confidence." When she woke up, she could feel that something had changed inside her. "It was like my anxiety was just turned off. I was ecstatic." Her therapist could scarcely believe her overnight transformation. "I came into his office, and he could just see that I was different. When I told him what I did, he almost fell out of his chair." Her new sense of composure lasted, and when it began to fade, she just repeated her mantra in her next lucid dream. She still suffers the occasional panic attack, but her anxiety has never returned in full force.

Sports scientists, meanwhile, have latched onto lucid dreaming as a tool in performance and exercise. In a series of experiments in the 2010s, Michael Schredl and Daniel Erlacher had lucid dreamers try to use their dreams to improve at physical tasks. In one study, forty people tried to toss a coin into a cup about six feet away. Afterward, one group was allowed to practice, another group tried to incubate lucid dreams about the coin toss, and a control group did nothing. When everyone attempted the task again, the people who had dreamed about it improved their hit rate by 43 percent, compared to just 4 percent for the control group. (Practicing while awake, though, was the most effective strategy.)

RECENT RESEARCH HAS vindicated much of Stephen's early work, but he is hardly bitter about the academic career he could have had. His books are still selling. His fans are ardent, his workshops well attended. Perhaps the spiritual experiences he has had in the dream state tempered his ambition. In one lucid dream, which he spent about half an hour recounting, he floated into a sky stippled with religious symbols and experienced a sense of oneness with the natural world as his body dissolved into a "point of awareness." He woke with his fear of death diminished. Lucid dreams have done enough for him.

EPILOGUE: MY NIGHT LIFE

I'VE BEEN KEEPING A DREAM JOURNAL EVER SINCE that summer in Peru; it's the only kind of diary I've ever kept. Whenever I've attempted to write in a traditional journal, I become stilted and self-conscious; with no structure to follow, I freeze up. But transcribing my dreams is easy. I've trained myself to focus on them as soon as I wake up—to rescue an image or a feeling from the haze of morning, to follow a thread until I can recall a whole story or scene. The entries in my dream journal have gotten longer and more detailed. They've grown from brief scenes and impressions into full narratives almost every time I wake up. I like seeing proof that even while I've been unconscious, I've been alive, feeling, doing. No matter how great or small the impact of dreams on daily life, I care about them because I experience them; even if I forget them, they are real in the moment.

Flipping through the journal—as I do from time to time—allows me not only to reenter the strange worlds I've invented in my sleep, but to remember the circumstances that inspired them. In senior year of college, I see anxiety about the future creeping in. In one dream, I'm ambling through a grassy field with a group of friends when we come across a family of seals who turn into men in suits and urge us to become management consultants. The seed of anxiety that transformed the sea mammals into businessmen is probably the same one responsible for my halfhearted (and never submitted) application to Credit Suisse.

Around finals, there's a spate of classic exam nightmares, as well as one in which I notify my professors that I've changed my major to geomorphology. Later, after I finally snagged a job writing (often personal) essays for the internet, I had a nightmare about a tabloid journalist snapping photos of me on the toilet. I imagine that my mind was helping me work out that line between what I was comfortable revealing to the world and what was too much.

When I want to incubate a lucid dream, I set some time aside. I pick a week when I don't expect any major life events or stressful changes. I'm extra-diligent about my dream journal. I wear the bracelet from Hawaii, the bright blue one that tells me to awaken in my dreams, and I use it for reality tests as often as I can remember. I use a meditation app, Headspace, on my phone. I don't set alarms for the middle of the night, but I often wake up in the early hours anyway, and if I'm trying to lucid dream, I don't panic over finding myself awake at four a.m. Instead, I appreciate the chance to focus on my intention to become lucid when I fall back to sleep. One of the pleasures of learning to lucid dream—or even of just trying to do so—is the way the training exercises sharpen and even recalibrate the experience of reality. Those middle-of-the-night hours are imbued with a sense of opportunity instead of anxiety.

Knowing what I know now, I trust my dreams more. I trust that they have continuity not only with waking behavior, but with semiconscious thoughts and fantasies. I believe that my dreams have helped me come to terms with romantic relationships and clarify feelings about friendship. Recently, I noticed that an old friend I'd lost touch with—and convinced myself I didn't want to reconnect with—kept popping up in my dreams, often just hovering in the background, making her presence felt in incongruous situations. Here she is in my ballet class. There she is at a company Christmas party. I thought about these dreams during the day, and I realized that I wasn't as over the friendship as I'd thought. I decided to text her, and we repaired our relationship.

As much as we have learned about dreams over the past couple of decades, the pace of research could surge if any one of a number of new technologies pan out. Scientists are still criticized for relying on their subjects to describe their own dreams; functional imaging can show which areas of the brain are active during sleep, but it's never been possible to know if people were honest about the details of their dreams. That might change soon. In 2013, a team of Japanese researchers led by neuroscientist Tomoyasu Horikawa published the results of a study in which they tried to decode the content of their subjects' dreams in real time. Horikawa had three young men take several naps inside an fMRI scanner while they were also hooked up to an EEG machine. When he saw that they were in REM, he would wake them and ask them to describe their dreams. After collecting at least two hundred dream snippets from each volunteer, he compiled a list of motifs that came up most often — things like cars, computers, books, women, and men. Next, Horikawa monitored the men's brain activity as they looked at images of those elements and then used this data to draw up a crude electronic dream dictionary of sorts — correlating specific fMRI patterns with elements of dreams. When the men went to sleep again, Horikawa watched the fMRI scans and guessed which of the motifs they were dreaming about. With a surprising degree of accuracy, the algorithm's guesses matched the men's own dream reports. It wasn't exactly a dream reader — it couldn't say which man or woman or book or car the person was dreaming about or what his emotions about it were — but it could be on the path to becoming one.

In the meantime, Matt Wilson predicts that researchers will continue to rely on rats. "I think the answers will come from rodent models," he said. He's most excited about the possibility of manipulating memories through dreams, as Gaetan de Lavilléon found was possible among mice. "We're interested in trying to influence dream content at a very detailed level — to create potentially new content during dreams. And there's this idea of selective learning — that is, pairing manipulations of reward signaling." In theory, "you could control the learning process if you could get rats to learn specific things by either manipulating dream content or by selectively reinforcing certain dream content."

When I started working on this book, I worried that understanding too much about dreams might detract from the mystery, the element of wonder that drew me to this subject in the first place. I am happy to report that, as much as I have learned about how my brain builds new worlds while I'm unconscious, the fundamental weirdness of dreams is as delightful and, in many ways, as enigmatic as ever. Just as knowing that a dopamine surge is involved in falling in love doesn't lessen the exhilaration, neither does knowing the neural correlates of dreams diminish our pleasure or alarm over their memories. Few people know as much about the biological mechanisms behind dreaming as William Dement, but even he quit smoking after dreaming of lung cancer. Allan Hobson — the anti-Freud himself — not only kept detailed dream journals for decades; he even judged his own dreams sufficiently interesting to publish. In *Thirteen Dreams Freud Never Had,* Hobson used his dreams to elucidate the biological processes involved in dreaming as well as the emotional arc of his life. He doesn't see those projects as incompatible.

Even as I can recognize patterns and themes in my dreams, I am still constantly surprised by the specific images and story lines my brain comes up with. We can locate some degree of order amid the chaos, but we are far from knowing why a certain memory mixes with another, or why our brains choose a certain night to play a particular scene. And the allure of dreams lives in the impossibility of ever fully untangling them. One night recently, I babysat a bodiless Japanese infant who spoke only Spanish. On another, I visited the writer Janet Malcolm at a nursing home for retired mathematicians and asked for her advice on how to structure my book. Some of my dreams don't inspire anything or bring buried memories to the surface or even make a whole lot of sense. They're escapism, allowing me to transcend the mundane reality of digging or studying. They're entertainment. They are evidence that I've lived in my sleep.

I vividly remember the first time I saw a human brain; it was floating in a jar of formalin up on a high shelf in my ninth-grade science classroom. One day after school, I lingered after everyone else had left, balanced a stool on top of a table, climbed up, and gaped at the brain. How could this little nugget of squishy-looking tubes control everything that made me who I was? I was amazed; I started telling people I was going to become a neuroscientist. That fantasy didn't last long, and I hadn't thought about that brain in years, but in writing this book, I've often felt that sense of awe I'd almost forgotten.

Psychologist Rubin Naiman has argued that the loss of dreams in our culture constitutes a bona fide public health hazard. Recognizing the importance of dreams for mental and cognitive health is especially crucial now, when pills that interfere with REM sleep are so often prescribed. Popular medications, including opioids, benzodiazepines, and some antidepressants, are known to suppress dreams. So let's start talking about dreams. Let's treat them like the real and profound experiences they are. Let's give them their rightful place in the world.

ACKNOWLEDGMENTS

This book has been very much a group project.

I am grateful above all to my editor, Eamon Dolan, whose optimism never failed to buoy me up and who has an uncanny ability to find structure where there is none, and to the entire team at Houghton Mifflin Harcourt, especially Tracy Roe, Lisa Glover, Rosemary McGuinness, Michael Dudding, Stephanie Buschardt, and Deb Brody. I'd also like to thank the following people:

My agent Bridget Wagner Matzie, who believed I could write a book before I believed it myself; and the rest of the team at Aevitas, including Chelsey Heller, Erica Bauman, and Elias Altman.

Gillian Brassil and Jane Hu, for superb fact-checking and incredible patience. I am inspired by your tolerance for the nitty-gritty.

The scientists who generously made time to explain their work and welcomed me into their labs, especially Patrick McNamara, Robert Stickgold, Matt Wilson, Stephen LaBerge, Deirdre Barrett, Hannah Wirtshafter, Mark Blechner.

The writers whose books were invaluable in my own research: Kelly Bulkeley, Jennifer Windt, William Domhoff, Andrea Rock.

The friends who read not-yet-coherent chapters and drafts, who knew when I needed feedback and when I needed encouragement, and who debated individual sentences until the last possible minute (even on vacation): Adam Plunkett, Esther Breger, Julia Fisher, Claire Groden, Isaac Chotiner, Lane Florsheim, Genevieve Walker, Jesse Singal, Emily Holleman, Brooke Shuman, Emily Fry, Alex Stone, Melissa Dahl, Meredith Turits, Justin Elliott.

Mom, champion copyeditor and cheerleader, and Dad, my most enthusiastic publicist.

The members of my dream group, who reminded me every month that people do, in fact, care about dreams.

And my dear friend James Rowland, who inspired this project all those years ago.

NOTES

Introduction

page

2 *"Proverbially, and undeniably":* Stephen LaBerge and Howard Rheingold, *Exploring the World of Lucid Dreaming* (New York: Ballantine, 1990), 9.

10 to 20 percent: Jayne Gackenbach, "An Estimate of Lucid Dreaming Incidence," *Lucidity* 10 (1991): 231; Michael Schredl and Daniel Erlacher, "Frequency of Lucid Dreaming in a Representative German Sample," *Perceptual and Motor Skills* 112, no. 1 (2011): 104–8, doi: 10.2466/09.PMS.112.1.104-108.

orgasm in one-third: LaBerge and Rheingold, *Exploring the World of Lucid Dreaming*, 171.

4 *ancient Greece:* William V. Harris, *Dreams and Experience in Classical Antiquity* (Cambridge, MA: Harvard University Press, 2009), 91.

medieval Japan: David Burton, *Buddhism: A Contemporary Philosophical Investigation* (London: Routledge, 2017), 103.

the paralyzed can move: M. Saurat et al., "Walking Dreams in Congenital and Acquired Paraplegia," *Consciousness and Cognition* 20, no. 4 (2011): 1425–32, doi: 10.1016/j.concog.2011.05.015.

5 *Gandhi argued:* Uma Majmudar, *Gandhi's Pilgrimage of Faith: From Darkness to Light* (Albany: State University of New York Press, 2005), 171.

Osama bin Laden: Iain R. Edgar, *The Dream in Islam: From Qur'anic Tradition to Jihadist Inspiration* (New York: Berghahn, 2011), 66–70.

less than 3 percent: Patrick McNamara, "People Who Don't Dream Might Not Recall Their Dreams," *Psychology Today,* October 25, 2015, https://www.psychologytoday.com/blog/dream-catcher/201510/people-who-dont-dream-might-not-recall-their-dreams.

"If you must sleep through a third of your life": Stephen LaBerge, *Lucid Dreaming* (Los Angeles: J. P. Tarcher, 1985), e-text.

6 *paper on extrasensory perception:* Simon Sherwood et al., "Dream Clairvoyance Study II Using Dynamic Video-Clips: Investigation of Consensus Voting Judging Procedures and Target Emotionality," *Dreaming* 10, no. 4 (2000): 221–36.

four hundred in 1998: "American Academy of Sleep Medicine Accreditation: Facts 06 04 08," https://aasm.org/resources/factsheets/accreditation.pdf.

more than twenty-five hundred today: Lynn Celmer, "Demand for Treatment of Sleep Illness Is Up as Drowsy Americans Seek Help for Potentially Dangerous Conditions," American Academy of Sleep Medicine, December 19, 2012, https://aasm.org/demand-for-treatment-of-sleep-illness-is-up-as-drowsy-americans-seek-help-for-potentially-dangerous-conditions/.

more than fifty billion: "$79.85 Billion Sleep Aids Market by Product and Sleep Disorder — Global Opportunity Analysis and Industry Forecast, 2017–2023 — Research and Markets," *BusinessWire,* May 5, 2017, https://www.businesswire.com/news/home/20170505005558/en/79.85-Billion-Sleep-Aids-Market-Product-Sleep.

7 *Philosophers:* Jennifer M. Windt, *Dreaming: A Conceptual Framework for Philosophy of Mind and Empirical Research* (Cambridge, MA: MIT Press, 2015).

1. How We Forgot About Dreams

11 *Muhammad's own dreams:* Kelly Bulkeley, *Dreaming in the World's Religions: A Comparative History* (New York: NYU Press, 2008), 193.

Hindu scripture teaches: Ibid., 27.

12 *"Beginnings of diseases":* Aristotle as quoted in S. M. Oberhelman, "Galen

on Diagnosis from Dreams," *Journal of the History of Medicine and Allied Sciences* 38 (1985): 37.

Hippocrates: Naphtali Lewis, *The Interpretation of Dreams and Portents in Antiquity* (Mundelein, IL: Bolchazy-Carducci, 1976), 30.

"have saved many people": Galen quoted in Edward Tick, *The Practice of Dream Healing: Bringing Ancient Greek Mysteries into Modern Medicine* (Wheaton, IL: Quest Books, 2001), 127.

One inscription tells: Ibid., 105.

13 *Methodist missionary once complained:* Jonathan W. White, *Midnight in America: Darkness, Sleep, and Dreams During the Civil War* (Chapel Hill: University of North Carolina Press, 2017), 98.

Some scholars even argue: Patrick McNamara and Kelly Bulkeley, "Dreams as a Source of Supernatural Agent Concepts," *Frontiers in Psychology* 6 (2015): 283, doi: 10.3389/fpsyg.2015.00283.

psychologists Richard Schweickert and Zhuangzhuang Xi: Richard Schweickert and Zhuangzhuang Xi, "Metamorphosed Characters in Dreams: Constraints of Conceptual Structure and Amount of Theory of Mind," *Cognitive Science* 34, no. 4 (2010): 665–84, doi: 10.1111/j.1551-6709.2009.01082.x.

14 *McNamara wrote:* Patrick McNamara, "Dreams and Revelations," *Aeon*, September 5, 2016.

15 *historian Andrew Burstein wrote:* Andrew Burstein, *Lincoln Dreamt He Died* (New York: St. Martin's Press, 2013), 48.

16 *French physician Louis Alfred Maury:* "Pre-History: Hypotheses About Dreams from Ancient Times to the End of the 19th Century," Freud Museum in London, https://www.freud.org.uk/education/topic/10576/subtopic/40021/.

developmental biologist Charles Child: Deirdre Barrett, *The Committee of Sleep: How Artists, Scientists, and Athletes Use Their Dreams for Creative Problem Solving—and How You Can Too* (New York: Crown, 2001), 163.

17 The Interpretation of Dreams: Sigmund Freud, *The Interpretation of Dreams,* trans. A. A. Brill (London: George Allen and Unwin, 1913).

18 *Milan Kundera mused:* Milan Kundera, *Identity,* trans. Linda Asher (New York: Harper Perennial, 1997), 5.

20 *"eldest son":* Quoted in Robert Van de Castle, *Our Dreaming Mind* (New York: Ballantine, 1994), 141.

"rests upon a deeper layer": Carl Jung, *The Archetypes and the Collective Unconscious,* 2nd ed., trans. R.F.C. Hull (Princeton, NJ: Princeton University Press, 1981), 3.

21 *One patient of his:* Carl Jung, *Modern Man in Search of a Soul,* trans. W. S. Dell and Cary F. Baynes (Orlando: Harcourt, 1933), 19–21.

22 *American Indian arts and crafts:* Philip Jenkins, *Dream Catchers: How Mainstream America Discovered Native Spirituality* (Oxford: Oxford University Press, 2004), 18, 96.

Jacques Frémin, wrote: Quoted in Anthony F. C. Wallace, "Dreams and the Wishes of the Soul: A Type of Psychoanalytic Theory Among the Seventeenth Century Iroquois," *American Anthropologist* 60, no. 2 (1958): 235.

chop off his own finger: Francesco Gioseppe Bressani, *The Jesuit Relations and Allied Documents,* vol. 39, ed. Reuben Gold Thwaites (Cleveland: Burrows Brothers, 1899), 18.

"They have a faith in dreams": Ibid., vol. 10, 169.

23 *anthropologist Jackson Steward Lincoln:* Jackson Steward Lincoln, *The Dream in Native American and Other Primitive Cultures* (Mineola, NY: Dover, 2003), 207–57.

"The ritual was deliberately designed": Bulkeley, *Dreaming in the World's Religions,* 261.

"The dream is an actual experience": Michele Stephen, "'Dreaming Is Another Power!': The Social Significance of Dreams Among the Mekeo of Papua New Guinea," *Oceania* 53, no. 2 (1982): 106–22.

24 *Sylvie Poirier:* Sylvie Poirier, "This Is Good Country. We Are Good Dreamers," in *Dream Travelers: Sleep Experiences and Culture in the Western Pacific,* ed. Roger Ivar Lohmann (New York: Palgrave Macmillan, 2003), 107–25.

the Rarámuri of northwestern Mexico: William Merrill, "The Rarámuri Stereotype of Dreams," in *Dreaming: Anthropological and Psychological Interpretations,* ed. Barbara Tedlock (Santa Fe: School of American Research Press, 1992), 194, 203.

25 *collecting dreams from their students at Case Western:* William G. Domhoff, "A Brief Biography of Calvin S. Hall," https://www2.ucsc.edu/dreams/About/calvin.html.

Aggressive encounters outnumbered friendly ones: Calvin Hall and William Domhoff, "Friendliness in Dreams," *Journal of Social Psychology* 62, no. 2 (1964): 309.

Sex appeared in men's dreams: William Domhoff, "The Dreams of Men and Women: Patterns of Gender Similarity and Difference" (2005), https://www2.ucsc.edu/dreams/Library/domhoff_2005c.html.

26 *characters in schizophrenic dreams:* William Domhoff, "The Content of Dreams: Methodologic and Theoretical Implications," in *Principles and Practices of Sleep Medicine,* 4th ed., ed. Meir Kryger et al. (Philadelphia: W. B. Saunders, 2005), 522–34.

"If you think about the social situation": Author's interview with Milton Kramer, July 6, 2017.

27 *questioning Freudian theory:* William Domhoff, "Moving Dream Theory Beyond Freud and Jung," paper presented at the symposium "Beyond Freud and Jung?," Graduate Theological Union, Berkeley, California, September 23, 2000, https://www2.ucsc.edu/dreams/Library/domhoff_2000d.html.

psychologist David Foulkes: David Foulkes, *Children's Dreaming and the Development of Consciousness* (Cambridge, MA: Harvard University Press, 1999).

28 *Foulkes later told a journalist:* Andrea Rock, *The Mind at Night: The New Science of How and Why We Dream* (New York: Basic Books, 2004), 32.

The youngest kids in the study: Foulkes, *Children's Dreaming,* 57.

birds and calves: Ibid., 61.

When Dean was six: Ibid., 160.

29 *New York's Psychoanalytic Institute:* Ralph Greenson, "The Exceptional Position of the Dream in Psychoanalytic Practice," in *Essential Papers on Dreams,* ed. Melvin Lansky (New York: NYU Press, 1992), 84.

Allan Hobson cast: Rachel Aviv, "Hobson's Choice: Can Freud's Theory of Dream Hold Up Against Modern Neuroscience?," *Believer,* October 2007, https://www.believermag.com/issues/200710/?read=article_aviv.

30 *"arrogance":* J. Allan Hobson, *Dream Life: An Experimental Memoir* (Cambridge, MA: MIT Press, 2011), vii.

a novel theory of dreams: J. Allan Hobson and Robert McCarley, "The Brain as a Dream State Generator: An Activation-Synthesis Hypothesis of the Dream Process," *American Journal of Psychiatry* 134, no. 12 (1977): 1335–48.

31 *one exhausted participant:* J. Allan Hobson, *Thirteen Dreams Freud Never Had: The New Mind Science* (New York: Pi Press, 2005), 11.

fashionable to denounce: Bruce K. Alexander and Curtis P. Shelton, *A History of Psychology in Western Civilization* (Cambridge: Cambridge University Press, 2014), 410.

32 *Federal funding:* Rock, *The Mind at Night,* 39.

Stanford began running: John Edgar Coover, *Experiments in Psychical Research at Leland Stanford Junior University* (Palo Alto, CA: Stanford University Press, 1917), 31.

Duke University: Robert Franklin Durden, *The Launching of Duke University, 1924–1949* (Durham, NC: Duke University Press, 1993), 114.

communicate with each other in their sleep: Hornell Norris Hart, *The Enigma of Survival: The Case for and Against an After Life* (Springfield, IL: C. C. Thomas, 1959), 236.

Jon Kabat-Zinn: Jeff Wilson, *Mindful America: The Mutual Transformation of Buddhist Meditation and American Culture* (Oxford: Oxford University Press, 2014), 35.

33 *"their stubborn resistance":* Jenkins, *Dream Catchers,* 157.

"It is not surprising": Ibid., 18.

"Through American history": Ibid., 154.

elaborate rules: Montague Ullman, "Dream Telepathy—Experimental and Clinical Findings," in *Lands of Darkness: Psychoanalysis and the Paranormal,* ed. Nick Totton (London: Karnac Books, 2003).

More than 60 percent of the time: Stanley Krippner, "Anomalous Experiences and Dreams," in *The New Science of Dreaming,* vol. 2, eds. Deirdre Barrett and Patrick McNamara (Westport, CT: Praeger Perspectives, 2007), 295.

one of their most famous studies: Montague Ullman and Stanley Krippner,

Dream Telepathy: Experiments in Nocturnal Extrasensory Perception (Newburyport, MA: Hampton Roads, 2003), 171.

34 *Robert Van de Castle spent eight nights:* Van de Castle, *Our Dreaming Mind,* 416–19.

"In retrospect, I am surprised": Author's interview with Stanley Krippner, April 13, 2016.

Maimonides team won a grant: Gordon T. Thompson, "Federal Grant Supports ESP Dream Research at Maimonides," *New York Times,* November 25, 1973, https://www.nytimes.com/1973/11/25/archives/federal-grant-supports-esp-dream-reserach-at-maimonides.html.

The first "premonitions bureau": Richard Wiseman, *Paranormality: Why We See What Isn't There* (N.P.: Spin Solutions, 2010), 144–46.

35 *seventy-six people wrote back:* John C. Barker, "Premonitions of the Aberfan Disaster," *Journal of the Society for Psychical Research* 44 (1967): 168–81.

novelist Vladimir Nabokov: Gennady Barabtarlo, *Insomniac Dreams: Experiments with Time by Vladimir Nabokov* (Princeton, NJ: Princeton University Press, 2017).

36 *"If there are associative patterns":* Sue Llewellyn, "Are Dreams Predictions?," *Aeon,* May 23, 2016, https://aeon.co/essays/how-dreams-predict-the-future-by-making-sense-of-the-past.

37 *"For months I dreamed about it, literally":* "Paul Tillich Dies; Theologian Was 79," *New York Times,* October 23, 1965, https://timesmachine.nytimes.com/timesmachine/1965/10/23/96719832.html?pageNumber=1.

German journalist Charlotte Beradt: Charlotte Beradt, *The Third Reich of Dreams* (Chicago: Quadrangle Books, 1968).

38 *Bruno Bettelheim noted:* Bruno Bettelheim, afterword in ibid.

"You have lots of dreams and encounter lots of events": Wiseman, *Paranormality,* 150.

39 *tried to replicate:* Edward Belvedere and David Foulkes, "Telepathy and Dreams: A Failure to Replicate," *Perceptual and Motor Skills* 33, no. 3 (1971): 783, doi: 10.2466/pms.1971.33.3.783.

Arthur Koestler: Caroline Watt, "Twenty Years at the Koestler Parapsychology Unit," *Psychologist* 19 (2006): 424–27.

video resembled their dreams: Caroline Watt, "Precognitive Dreaming:

Investigating Anomalous Cognition and Psychological Factors," *Journal of Parapsychology* 78, no. 1 (2014): 115–25.

selective memory: Caroline Watt et al., "Psychological Factors in Precognitive Dream Experiences: The Role of Paranormal Belief, Selective Recall and Propensity to Find Correspondences," *International Journal of Dream Research* 7, no. 1 (2014): 1–8, doi: 10.11588/ijodr.2014.1.11218.

2. The Vanguard

41 *Eugene Aserinsky:* Chip Brown, "The Stubborn Scientist Who Unraveled a Mystery of the Night," *Smithsonian,* October 2003, https://www.smithsonianmag.com/science-nature/the-stubborn-scientist-who-unraveled-a-mystery-of-the-night-91514538/.

"All kinds of things had to be rigged": Author's interview with Armond Aserinsky, July 14, 2017.

42 *drafted into the army:* Lynne Lamberg, "The Student, the Professor and the Birth of Modern Sleep Research," *Medicine on the Midway* (Spring 2004), http://www.uchospitals.edu/pdf/uch_006319.pdf.

There was "no joy": Eugene Aserinsky, "Memories of Famous Neuropsychologists: The Discovery of REM," *Journal of the History of the Neurosciences* 5, no. 3 (1996): 213.

43 *month underground in a cave:* Anna Azvolinsky, "Cave Dwellers, 1938," *Scientist* (March 2016), https://www.the-scientist.com/?articles.view/articleNo/45359/title/Cave-Dwellers—1938/.

"I was forewarned": Aserinsky, "Memories of Famous Neuropsychologists," 214.

45 Science *in 1953:* Eugene Aserinsky and Nathaniel Kleitman, "Regularly Occurring Periods of Eye Motility, and Concomitant Phenomena, During Sleep," *Science* 118, no. 3062 (1953): 273.

46 *cameo in the comedy* Sleepwalk with Me: Kate Murphy, "Catching Up on Sleep with Dr. William C. Dement," *New York Times,* September 22, 2012, http://www.nytimes.com/2012/09/23/opinion/sunday/catching-up-on-sleep-with-dr-william-c-dement.html.

Born in 1947: Dorian Rolston, "The Dream Catcher," *Matter,* November 21, 2013, https://medium.com/matter/the-dream-catcher-c85e3bb29693.

"I was a complete introvert": Joan Libman, "Dr. Dreams: Stanford Scientist Stephen LaBerge Is Sleeping on the Idea that Dreams Are a Tool for Bettering Our Lives," *Los Angeles Times,* November 15, 1988, http://articles.latimes.com/1988-11-15/news/vw-221_1_lucid-dream/2.

47 *"I would have the experience":* David Jay Brown, ed., *Mavericks of the Mind: Conversations for the New Millennium* (Emeryville, CA: Crossing Press, 1993).

"In Germany, for some reason": Author's interview with Stephen LaBerge, June 30, 2017.

"ground zero for the hippie movement": Author's interview with Stephen LaBerge, September 30, 2016.

48 *"the explanation was that I was dreaming":* LaBerge and Rheingold, *Exploring the World of Lucid Dreaming,* 67.

49 *self-generated illusions:* Bulkeley, *Dreaming in the World's Religions,* 100.

"emotional stress within the dream": Celia Green, *Lucid Dreams* (Oxford: Institute of Psychophysical Research, 1968), 30.

"Flying is a common feature": Ibid., 51.

"Persons who appear in lucid dreams": Ibid., 63.

"When one is asleep": Aristotle, "On Dreams," in LaBerge, *Lucid Dreaming.*

50 *"On this the youth inquired":* Philip Schaff, ed., *Nicene and Post-Nicene Fathers, First Series,* vol. 1, trans. J. G. Cunningham (Buffalo, NY: Christian Literature Publishing, 1887), 514.

Nietzsche described: Friedrich Nietzsche, quoted in LaBerge, *Lucid Dreaming.*

51 *"who are quite clearly aware":* Sigmund Freud, quoted in ibid.

aroused his "keenest interest": Frederik van Eeden, quoted in ibid.

"He who spends a third part of his life": Frederik van Eeden, *The Bride of Dreams,* trans. Mellie von Auw (New York: Mitchell Kennerley, 1913), 160.

"the orthodox view": LaBerge, *Lucid Dreaming.*

"Dream researchers": Author's interview with LaBerge, September 30, 2016.

52 *Dement tracked:* Howard Roffwarg et al., "Dream Imagery: Relationship

to Rapid Eye Movements of Sleep," *Archives of General Psychiatry* 7, no. 4 (1962): 235–58, doi: 10.1001/archpsyc.1962.01720040001001.

Ping-Pong game: William C. Dement, *Some Must Watch While Some Must Sleep: Exploring the World of Sleep* (New York: Norton, 1972), 118.

climbing a flight of five stairs: Roffwarg et al., "Dream Imagery."

53 *"My doctoral dissertation depended":* Stephen LaBerge, presentation at Dreaming and Awakening, Kalani Oceanside Retreat, Hawaii, September 25, 2016.

54 *fastened electrodes:* Stephen LaBerge et al., "Lucid Dreaming Verified by Volitional Communication During REM Sleep," *Perceptual and Motor Skills* 52 (1981): 727–32.

"It was a huge turnaround": Author's interview with Patricia Garfield, September 23, 2017.

"just drum circles": Author's interview with Erin Wamsley, November 5, 2015.

"It was something new": Author's interview with Stephen LaBerge, June 30, 2017.

55 *"The subjective accounts and physiological measures":* LaBerge et al., "Lucid Dreaming Verified," 731.

"I had to keep finding money": Author's interview with Stephen LaBerge, June 30, 2017.

56 *"skeptics who said things":* LaBerge, presentation at Dreaming and Awakening.

French doctor Louis Alfred Maury: Jacqueline Carroy, "Observer, Raconter ou Ressusciter Les Reves?," *Communications* 84 (2009): 139.

57 *brain specialization:* Stephen LaBerge and William Dement, "Lateralization of Alpha Activity for Dreamed Singing and Counting During REM Sleep," *Psychophysiology* 19 (1982): 331–32.

"soul- and body-shaking explosions": LaBerge and Rheingold, *Exploring the World of Lucid Dreaming,* 107.

58 *Miranda:* Stephen LaBerge et al., "Physiological Responses to Dreamed Sexual Activity During Lucid REM Sleep," *Psychophysiology* 20 (1993): 454–55.

59 *"Had I been able to get funding":* Libman, "Dr. Dreams."

3. Dreams Enter the Lab

60 *"The thing about natural behavior":* Author's interview with Matthew Wilson, March 16, 2017.

61 *type of neuron:* Johannes Niediek and Jonathan Bain, "Human Single-Unit Recordings Reveal a Link Between Place-Cells and Episodic Memory," *Frontiers in Systems Neuroscience* 8 (2014): 158.

"The first time you're exposed to an environment": Author's interview with Hannah Wirtshafter, April 5, 2017.

63 *his rodent research:* "Wilson Lab @ MIT," Massachusetts Institute of Technology, http://www.mit.edu/org/w/wilsonlab/html/publications.html.

London taxi drivers: E. A. Maguire et al., "London Taxi Drivers and Bus Drivers: A Structural MRI and Neuropsychological Analysis," *Hippocampus* 16, no. 12 (2006): 1091.

Harvard psychiatrist Robert Stickgold: Robert Stickgold et al., "Replaying the Game: Hypnagogic Images in Normals and Amnesics," *Science* 290, no. 5490 (2000): 350–53, doi: 10.1126/science.290.5490.350.

64 *"I had been staying up in Vermont":* Author's interview with Robert Stickgold, March 13, 2017.

66 *"They were just really nice":* Author's interview with David Roddenberry, March 22, 2017.

67 *"I was always translating calculations":* Author's interview with Joseph De Koninck, March 30, 2017.

68 *recruiting groups of Anglophone:* Joseph De Koninck et al., "Intensive Language Learning and Increases in Rapid Eye Movement Sleep: Evidence of a Performance Factor," *International Journal of Psychophysiology* 8, no. 1 (1989): 43–47.

tell him about their dreams: Joseph De Koninck et al., "Language Learning Efficiency, Dreams and REM Sleep," *Psychiatric Journal of the University of Ottawa* 15, no. 2 (1990): 91–92.

69 *"observed changes in dreams":* Joseph De Koninck et al., "Vertical Inversion of the Visual Field and REM Sleep Mentation," *Journal of Sleep Research* 5, no. 1 (1996): 16–20.

4. The Renaissance of Sleep Research

72 *check it out:* "Conferences and Events Archive," International Association for the Study of Dreams, http://www.asdreams.org/conferences-and-events-archive/.

recommended reading: "Books Reviewed in DreamTime," International Association for the Study of Dreams, http://www.asdreams.org/books-reviewed-in-dreamtime/.

Slavoj Zizek: Slavoj Zizek, *Trouble in Paradise: From the End of History to the End of Capitalism* (New York: Melville House, 2014).

73 *His dissertation:* "Lecturer Wins Prestigious International Award for Research into Lucid Dreams," University of Northampton, https://www.northampton.ac.uk/news/david-saunders-ernest-hartmann-award/.

74 *Neuroscientist Michael Schredl:* "Michael Schredl: Sleep Laboratory: Central Institute of Mental Health," Google Scholar, https://scholar.google.com/citations?hl=de&user=YNIpIfMAAAAJ&view_op=list_works&sortby=pubdate.

75 *anxiety:* Patricia Sagaspe et al., "Effects of Sleep Deprivation on Color-Word, Emotional and Specific Stroop Interference and on Self-Reported Anxiety," *Brain and Cognition* 60, no. 1 (2006): 76.

depression: Daniel E. Ford et al., "Epidemiological Study of Sleep Disturbances and Psychiatric Disorders: An Opportunity for Prevention?," *Journal of the American Medical Association* 262, no. 11 (1989): 1479.

heart disease: Najib Tayas et al., "A Prospective Study of Sleep Duration and Coronary Heart Disease in Women," *Journal of the American Medical Association Internal Medicine* 163, no. 2 (2003): 205.

weight gain: Sanjay R. Patel and Frank B. Hu, "Short Sleep Duration and Weight Gain: A Systematic Review," *Obesity* 16, no. 3 (2008): 643.

sleep deprivation and alcohol intoxication: A. M. Williamson and Anne-Marie Feyer, "Moderate Sleep Deprivation Produces Impairments in Cognitive and Motor Performance Equivalent to Legally Prescribed Levels of Alcohol Intoxication," *Occupational and Environmental Medicine* 57, no. 10 (2000): 649.

"Every hour": Matthew Walker, *Why We Sleep: Unlocking the Power of Sleep and Dreams* (New York: Scribner, 2017), 134.

medical residents: Steven W. Lockley et al., "Effect of Reducing Interns' Weekly Work Hours on Sleep and Attentional Failures," *New England Journal of Medicine* 351 (2004): 1829–37.

errors in diagnosis and prescription: Christopher P. Landrigan et al., "Effect of Reducing Interns' Work Hours on Serious Medical Errors in Intensive Care Units," *New England Journal of Medicine* 351 (2004): 1838.

glymphatic system: N. A. Jessen et al., "The Glymphatic System: A Beginner's Guide," *Neurochemical Research* 40, no. 12 (2015): 2583.

accelerates: Andy R. Eugene and Jolanta Masiak, "The Neuroprotective Aspects of Sleep," *MEDtube Science* 3, no. 1 (2015): 35.

myelin: Michele Bellesi et al., "Effects of Sleep and Wake on Oligodendrocytes and Their Precursors," *Journal of Neuroscience* 33, no. 36 (2013): 14288, doi: 10.1523/JNEUROSCI.5102-12.2013.

76 *Human growth hormone:* E. Van Cauter and L. Plat, "Physiology of Growth Hormone During Sleep," *Journal of Pediatrics* 128, no. 5 (1996): S32.

weaken the immune system: Alexandros N. Vgontzas, "Sleep Deprivation Effects on the Activity of the Hypothalamic-Pituitary-Adrenal and Growth Axes: Potential Clinical Implications," *Clinical Endocrinology* 51 (1999), doi: 10.1046/j.1365-2265.1999.00763.x.

risk factor for hypertension: "Sleep and Disease Risk," Division of Sleep Medicine at Harvard Medical School, http://healthysleep.med.harvard.edu/healthy/matters/consequences/sleep-and-disease-risk.

appetite haywire: Ibid.

nearly five hundred adults: G. Halser et al., "The Association Between Short Sleep Duration and Obesity in Young Adults: A 13-Year Prospective Study," *Sleep* 27, no. 4 (2004): 661.

slept for less than five hours: D. J. Gottlieb et al., "Association of Sleep Time with Diabetes Mellitus and Impaired Glucose Tolerance," *Archives of Internal Medicine* 165, no. 8 (2005): 863.

look at upsetting images: Els van der Helm et al., "REM Sleep De-

Potentiates Amygdala Activity to Previous Emotional Experiences," *Current Biology* 21, no. 23 (2011): 2029, doi: 10.1016/j.cub.2011.10.052.

irritability, paranoia, anger: Annie Gordon, "Up All Night: The Effects of Sleep Loss on Mood," *Psychology Today,* August 15, 2013, https://www.psychologytoday.com/blog/between-you-and-me/201308/all-night-the-effects-sleep-loss-mood.

"My body had no more feeling": Haruki Murakami, "Sleep," trans. Jay Rubin, *New Yorker,* March 30, 1992, 34.

January of 1959: Gay Gaer Luce, "Sleep Deprivation," *Current Research on Sleep and Dreams* (Washington, DC: Public Health Service Publication no. 1389, 1973).

77 *"tired but normal":* "Stay-Awake Man Half Way to Goal," *New York Times,* January 25, 1959.

better part of that time in REM: Geoff Rolls, *Classic Case Studies in Psychology,* 3rd ed. (London: Routledge, 2015), 251–52.

78 *"We did it carefully":* "Secrets of Sleep — Sleep Deprivation — Peter Tripp Part 2/2," YouTube video, posted November 26, 2012, https://www.youtube.com/watch?v=2tlsB0OXz4E.

in 1965, Randy Gardner: Shankar Vedantam, "The Haunting Effects of Going Without Sleep," *Morning Edition,* NPR, December 27, 2017.

resolved to break: Rolls, *Classic Case Studies,* 253–56.

Guinness Book of World Records: Michael Horsnell, "Man Who Stayed Up for 266 Hours Awakes to Bad News," *Irish Independent,* May 26, 2007, https://www.independent.ie/world-news/man-who-stayed-up-for-266-hours-awakes-to-bad-news-about-the-record-26293322.html.

79 *Lapses in memory:* Robbert Havekes et al., "Sleep Deprivation Causes Memory Deficits by Negatively Impacting Neuronal Connectivity in Hippocampal Area CA1," *eLife* 5 (2016): 13424, doi: 10.7554/eLife.13424.

can wreck our ability: Paula Alhola and Paivi Polo-Kantola, "Sleep Deprivation: Impact on Cognitive Performance," *Neuropsychiatric Disease and Treatment* 3, no. 5 (2007): 553.

navigating a virtual maze: Nam Nguyen et al., "Overnight Sleep Enhances Hippocampus-Dependent Aspects of Spatial Memory," *Sleep* 36, no. 7 (2013): 1051, doi: 10.5665/sleep.2808.

mastering physical tasks: Yusuf Patrick et al., "Effects of Sleep Deprivation on Cognitive and Physical Performance in University Students," *Sleep and Biological Rhythms* 15, no. 3 (2017): 217, doi: 10.1007/s41105-017-0099-5.

poor grades: Shelley Hershner and Ronald Chervin, "Causes and Consequences of Sleepiness Among College Students," *Nature and Science of Sleep* 6 (2014): 73, doi: 10.2147/NSS.S62907.

low scores: Eric Ride and Mark Showalter, "Sleep and Student Achievement," *Eastern Economic Journal* 38, no. 4 (2012): 512.

miss out on REM: Christie Nicholson, "Strange but True: Less Sleep Means More Dreams," *Scientific American,* September 20, 2007, https://www.scientificamerican.com/article/strange-but-true-less-sleep-means-more-dreams/.

rats die: A. Rechtschaffen and B. M. Bergmann, "Sleep Deprivation in the Rat by the Disk-Over-Water Method," *Behavioral Brain Research* 69, no. 2 (1995): 55.

they grew reckless: William Dement, "The Paradox of Sleep: The Early Years," *Archives Italiennes de Biologie* 142 (2004): 340–41.

80 *giving out eye masks and earplugs:* Heather Schofield, "Development and Behavioral Economics Lab in Chennai, India," Center for Global Health at the Perelman School of Medicine-University of Pennsylvania, https://www.med.upenn.edu/globalhealth/development-and-behavioral-economics-lab-in-chennai-india.html.

5. Problem-Solving

82 *work through real issues:* Barrett, *The Committee of Sleep,* 164–66.

83 *brainteasers and instructions:* Ibid., 163–64.

84 *Morton Schatzman:* Ibid., 170–75.

85 *"mental doodlings":* Morton Schatzman, "The Meaning of Dreaming," *New Scientist,* December 25, 1986.

"This is a network of brain regions": Author's interview with Robert Stickgold, March 13, 2017.

discovered by accident: Randy L. Buckner et al., "The Brain's Default Network: Anatomy, Function, and Relevance to Disease," *Annals of the New York Academy of Sciences* 1124, no. 1 (2008): 2.

86 *mind-wandering:* A. Sood and D. T. Jones, "On Mind Wandering, Attention, Brain Networks, and Meditation," *Explore* 9, no. 3 (2013): 136.

creative thinking: Roger E. Beaty et al., "Creativity and the Default Network: A Functional Connectivity Analysis of the Creative Brain at Rest," *Neuropsychologia* 64 (2014): 92, doi: 10.1016/j.neuropsychologia.2014.09.019.

Kieran Fox: Kieran C. R. Fox et al., "Dreaming as Mind Wandering: Evidence from Functional Neuroimaging and First-Person Content Reports," *Frontiers in Human Neuroscience* 7 (2013): 412, doi: 10.3389/fnhum.2013.00412.

87 *looser, less obvious word associations:* Robert Stickgold et al., "Sleep-Induced Changes in Associative Memory," *Journal of Cognitive Neuroscience* 11, no. 2 (1999): 182–93.

88 *Psychologist Ernest Schachtel:* Ernest Schachtel, *Metamorphosis: On the Conflict of Human Development and the Psychology of Creativity* (New York: Basic Books, 1959), 308.

Men tend: Michael Schredl and Iris Reinhard, "Gender Differences in Dream Recall: A Meta-Analysis," *Journal of Sleep Research* 17, no. 2 (2008): 125, doi: 10.1111/j.1365-2869.2008.00626.x.

as do older people: Tore Nielsen, "Variations in Dream Recall Frequency and Dream Theme Diversity by Age and Sex," *Frontiers in Neurology* 3 (2012): 106, doi: 10.3389/fneur.2012.00106.

89 *"openness to experience":* Michael Schredl et al., "Dream Recall Frequency, Attitude Towards Dreams and Openness to Experience," *Dreaming* 13, no. 3 (2003): 145, doi: https://doi.org/10.1023/A:1025369311813.

"tolerance of ambiguity": J. Houran and R. Lange, "Modeling Precognitive Dreams as Meaningful Coincidences," *Psychological Reports* 83, no. 3, pt. 2 (1998): 1411, doi: 10.2466/pr0.1998.83.3f.1411.

stave off dementia: Joshua Jackson et al., "Can an Old Dog Learn (and Want to Experience) New Tricks? Cognitive Training Increases Openness to Experience in Older Adults," *Psychology and Aging* 27, no. 2 (2012): 286, doi: 10.1037/a0025918.

"There are certain models": Diana Yates, "Enhancing Cognition in Older Adults Also Changes Personality," University of Illinois News Bureau, January 18, 2012, https://news.illinois.edu/view/6367/205159.

Jeremy Taylor: Jeremy Taylor, *The Wisdom of Your Dreams: Using Dreams to Tap into Your Unconscious and Transform Your Life* (New York: Penguin, 1992), 61.

alcohol suppresses REM: Maia Szalavitz, "Sleeping It Off: How Alcohol Affects Sleep Quality," *Healthland* (blog), *Time*, February 8, 2013, http://healthland.time.com/2013/02/08/sleeping-it-off-how-alcohol-affects-sleep-quality/.

90 *vitamin B_6:* Matthew Ebben et al., "Effects of Pyridoxine on Dreaming: A Preliminary Study," *Perceptual and Motor Skills* 94, no. 1 (2002): 135, doi: https://doi.org/10.2466/pms.2002.94.1.135.

Meg Jay advises: Author's interview with Meg Jay, July 6, 2017.

psychologist Henry Reed: Henry Reed, "Learning to Remember Dreams," *Journal of Humanistic Psychology* 13, no. 3 (1973): 33–48, doi: 10.1177/002216787301300305.

Amy Parke and Caroline Horton: Amy R. Parke and Caroline L. Horton, "A Re-Examination of the Interference Hypothesis on Dream Recall and Dream Salience," *International Journal of Dream Research* 2, no. 2 (2009): 60–63, doi: https://doi.org/10.11588/ijodr.2009.2.364.

91 *another strategy:* Taylor, *The Wisdom of Your Dreams*, 72.

"being abruptly ushered": Rubin Naiman, "Dreamless: The Silent Epidemic of REM Sleep Loss," *Annals of the New York Academy of Sciences* 1406 (2017): 80.

92 *the more time that elapses:* Deirdre Barrett, "What Processes in the Brain Allow You to Remember Dreams?," *Scientific American Mind* 25, June 12, 2014, https://www.scientificamerican.com/article/what-processes-in-the-brain-allow-you-to-remember-dreams/.

Dreaming is most frequent: Mark Solms, *The Brain and the Inner World: An Introduction to the Neuroscience of Subjective Experience* (New York: Other Press, 2002), 183.

93 *Beethoven:* Barrett, *The Committee of Sleep*, 68–69.

Paul McCartney: Ibid., 66–67.

Ingmar Bergman's: Barrett, *The Committee of Sleep*, 30.

Fellini's: Ibid.

Richard Linklater's: Rock, *The Mind at Night,* 147.

Mary Shelley: Barrett, *The Committee of Sleep,* 40.

E. B. White: Ibid., 46.

"Lascaux and Corvet": Kelly Bulkeley, *Big Dreams: The Science of Dreaming and the Origins of Religion* (Oxford: Oxford University Press, 2016), 83.

"The earliest documents": Michaela Schrage-Früh, *Philosophy, Dreaming and the Literary Imagination* (New York: Palgrave Macmillan, 2016), 55–56.

"openness to experience": Bulkeley, *Dreaming in the World's Religions,* 15.

94 *tendency to daydream:* David Watson, "To Dream, Perchance to Remember: Individual Differences in Dream Recall," *Personality and Individual Differences* 34, no. 7 (2003): 1271, doi: 10.1016/S0191-8869(02)00114-9.

physician James Pagel: James Pagel et al., "Dream Use in Filmmaking," *Dreaming* 9, no. 4 (1999): 247.

"They blew all the scales": Author's interview with James Pagel, March 29, 2017.

95 *Charles Simic:* Charles Simic, "Dreams I've Had (and Some I Haven't)," *Daily* (blog), *New York Review of Books,* January 24, 2013, http://www.nybooks.com/daily/2013/01/24/simic-dreams-had/.

Charlotte Brontë: Barrett, *The Committee of Sleep,* 43.

Salvador Dalí: Ibid., 5–6.

Robert Louis Stevenson: Ibid., 63.

Fifty Secrets of Magic Craftsmanship: Salvador Dalí, *Fifty Secrets of Magic Craftsmanship,* trans. Haakon M. Chevalier (Mineola, NY: Dover, 1948).

96 *target of bullies:* Maria Konnikova, "How to Beat Writer's Block," NewYorker.com, March 11, 2016, https://www.newyorker.com/science/maria-konnikova/how-to-beat-writers-block.

stomach full of aspirin: Kay Redfield Jamison, *Night Falls Fast: Understanding Suicide* (New York: Vintage, 2000), 98.

"an astonishing thing": Graham Greene, quoted in Norman Sherry, *The Life of Graham Greene, Volume One: 1904–1939* (New York: Penguin, 1989), 92.

"leaving his subconscious to work": Yvonne Cloetta, foreword to Graham Greene, *A World of My Own: A Dream Diary* (New York: Viking, 1992).

97 *"dreamed a lot of Sarah":* Graham Greene, *The End of the Affair* (New York: Penguin, 1951), 19.

"no sense of dizziness": Maya Angelou, quoted in Naomi Epel, *Writers Dreaming* (New York: Carol Southern Books, 1993), 26.

98 *Writer Kathryn Davis:* Patrick Lauppe, "Trial: A Conversation with Kathryn Davis," blog, *Harvard Advocate,* March 3, 2014, http://theharvardadvocate.com/blog/post/2014-3-7-trial-a-conversation-with-kathryn-davis/.

Stephen King: Epel, *Writers Dreaming,* 137–38.

99 *Otto Loewi:* Barrett, *The Committee of Sleep,* 90–92.

100 *"Easter Saturday":* Otto Loewi, *From the Workshop of Discoveries* (Lawrence: University of Kansas Press, 1953), 33.

101 *mathematician Donald Newman: A Brilliant Madness: John Nash,* directed by Mark Samels (PBS, 2002), documentary film transcript, https://cosmolearning.org/documentaries/a-brilliant-madness-john-nash-620/1/.

6. Preparation for Life

103 *threat-simulation hypothesis:* Antti Revonsuo, "The Reinterpretation of Dreams: An Evolutionary Hypothesis of the Function of Dreaming," *Behavioral and Brain Sciences* 23 (2000): 877.

104 *rats' ability to respond:* Dolores Martinez-Gonzalez et al., "REM Sleep Deprivation Induces Changes in Coping Responses that are not Reversed by Amphetamine," *Sleep* 27, no. 4 (2004): 609.

106 *researchers from the Sorbonne:* Isabelle Arnulf et al., "Will Students Pass a Competitive Exam that They Failed in Their Dreams?," *Consciousness and Cognition* 29 (2014): 36, doi: 10.1016/j.concog.2014.06.010.

"Negative anticipation": Ibid., 46.

memories of being rocked: Michael Schredl, "Personality Correlates of

Flying Dreams," *Imagination, Cognition and Personality* 27, no. 2 (2007): 135, doi: 10.2190/IC.27.2.d.

107 *Egyptian papyri:* Frederick L. Coolidge et al., *Dream Interpretation as a Psychotherapeutic Technique* (Boca Raton: CRC Press, 2006), 112.

Vedic scriptures: Ibid., 28.

Japanese college students: Richard Griffith et al., "The Universality of Typical Dreams: Japanese vs. Americans," *American Anthropologist* 60, no. 6 (1958): 1173, doi: 10.1525/aa.1958.60.6.02a00110.

21 percent of American undergraduates: Ibid., 1177.

Bar Hedya: Aron Moss, "What Does It Mean When You Dream Your Teeth Are Falling Out?," Chabad-Lubavitch Media Center, https://www.chabad.org/library/article_cdo/aid/2842585/jewish/What-Does-It-Mean-When-You-Dream-Your-Teeth-Are-Falling-Out.htmb.

the Navajo: Coolidge, *Dream Interpretation,* 115.

castration: Frank J. Sulloway, *Freud, Biologist of the Mind: Beyond the Psychoanalytic Legend* (Cambridge, MA: Harvard University Press, 1979), 344.

Sandor Lorand: Van de Castle, *Our Dreaming Mind,* 340.

fears aging: "Teeth Dreams," DreamDictionary.org, https://www.dreamdictionary.org/common/teeth-dreams/.

has misspoken: "Common Dreams: Teeth Dreams," Dream Moods, http://www.dreammoods.com/commondreams/teeth-dreams.html.

"some hardship": "Dreams About Teeth," DreamLookup.com, http://www.dreamlookup.com/index.php/search/level1/Teeth/.

empirical study: Frederick Coolidge, "The Loss of Teeth in Dreams: An Empirical Investigation," *Psychological Reports* 54, no. 3 (1984): 931–35, doi: 10.2466/pr0.1984.54.3.931.

108 *video game called Doom:* A. H. Pantoja et al., "Assessment of the Adaptive Value of Dreams," paper presented at the annual Society for Neuroscience Conference, Chicago, Illinois, October 19, 2009, http://www.abstractsonline.com/Plan/ViewAbstract.aspx?sKey=93f119e2-d112-481c-bf0e-c8e3a9ba665f&cKey=f1e4632c-97be-41ad-a1dd-7f5727445bc6&mKey=%7b081F7976-E4CD-4F3D-A0AF-E8387992A658%7d.1173.

109 *interview with* New Scientist: Ewen Callaway, "Dreams of Doom Help

Gamers Learn," *New Scientist,* October 30, 2009, https://www.newscientist.com/article/dn18082-dreams-of-doom-help-gamers-learn/.

Erin Wamsley: Erin Wamsley et al., "Dreaming of a Learning Task Is Associated with Enhanced Sleep-Dependent Memory Consolidation," *Current Biology* 20, no. 9 (2010): 850–55.

"If you give humans an engaging learning task": Author's interview with Erin Wamsley, November 6, 2015.

110 *hospital patients slated:* Louis Breger, Ian Hunter, and Ron Lane, *The Effect of Stress on Dreams* (New York: International Universities Press, 1971), 179.

111 *Brooklyn and Bethesda:* G. William Domhoff, "Realistic Simulation and Bizarreness in Dream Content: Past Findings and Suggestions for Future Research," in *The New Science of Dreaming: Content, Recall, and Personality Characteristics,* vol. 2 (Westport, CT: Praeger Press, 2007), 4–5.

constraints on the dreaming imagination: Robert Stickgold, Allan Hobson, and Cynthia Rittenhouse, "Constraint on the Transformation of Characters, Objects, and Settings in Dream Reports," *Consciousness and Cognition* 3, no. 1 (1994): 100.

112 *appropriate to the situation:* Antti Revonsuo and Christina Salmivalli, "A Content Analysis of Bizarre Elements in Dreams," *Dreaming* 5, no. 3 (1995): 169.

preteens with trouble sleeping: Maria M. Wong et al., "Childhood Sleep Problems, Response Inhibition, and Alcohol and Drug Outcomes in Adolescence and Young Adulthood," *Alcoholism: Clinical and Experimental Research* 34, no. 6 (2010): 1033.

people over sixty-five: C. L. Turvey et al., "Risk Factors for Late-Life Suicide: A Prospective, Community-Based Study," *American Journal of Geriatric Psychiatry* 10, no. 4 (2002): 398.

Matthew Walker and his team: Andrea N. Goldstein-Piekarski et al., "Sleep Deprivation Impairs the Human Central and Peripheral Nervous System Discrimination of Social Threat," *Journal of Neuroscience* 35, no. 28 (2015): 10135, doi: https://doi.org/10.1523/jneurosci.5254-14.2015.

113 *80 to 90 percent:* Anastasia Mangiaruga et al., "Spotlight on Dream

Recall: The Ages of Dreams," *Nature and Science of Sleep* 10 (2018): 1–12, doi: 10.2147/NSS.S135762.

severely depressed populations: Milton Kramer et al., "Depression: Dreams and Defenses," *American Journal of Psychiatry* 122, no. 4 (1965): 411–19.

shorter and less emotional: Deirdre Barrett and Michael Loeffler, "Comparison of Dream Content of Depressed vs. Nondepressed Dreamers," *Psychological Reports* 70, no. 2 (1992): 403–6.

fewer characters: Domhoff, "The Content of Dreams."

Writers on Depression: Nell Casey, ed., *Unholy Ghost: Writers on Depression* (New York: Harper Perennial, 2002).

Virginia Heffernan: Ibid., 9.

Lesley Dormen's: Ibid., 236.

"intolerable aspect of my illness": Styron in Epel, *Writers Dreaming*, 277.

114 *A healthy person typically:* "Natural Patterns of Sleep," Division of Sleep Medicine at Harvard Medical School and WGBH Educational Foundation, http://healthysleep.med.harvard.edu/healthy/science/what/sleep-patterns-rem-nrem.

Among the depressed: Maria Popova, "Dreaming, Depression, and How Sleep Affects Emotions," *Atlantic,* August 13, 2012, https://www.theatlantic.com/health/archive/2012/08/dreaming-depression-and-how-sleep-affects-emotions/261051/.

The emotional pattern of the night: Rosalind Cartwright et al., "Role of REM Sleep and Dream Variables in the Prediction of Remission from Depression," *Psychiatry Research* 80, no. 3 (1998): 249–255, doi: 10.1016/S0165-1781(98)00071-7.

"as much a pioneer": Walker, *Why We Sleep*, 211.

she invited sixty people: Rosalind Cartwright, "Dreams and Adaptation to Divorce," in *Trauma and Dreams,* ed. Deirdre Barrett (Cambridge, MA: Harvard University Press, 2001), 179–85.

115 *a closer look at the content:* Rosalind Cartwright et al., "Broken Dreams: A Study of the Effects of Divorce and Depression on Dream Content," *Psychiatry* 47, no. 3 (1984): 251–59, doi: 10.1080/00332747.1984.11024246.

three hundred mourners at a hospice center: Scott Wright et al., "The Impact of Dreams of the Deceased on Bereavement: A Survey of Hospice

Caregivers," *American Journal of Hospice and Palliative Medicine* 31, no. 2 (2014): 132, doi: 10.1177/1049909113479201.

116 *"'Don't tell me your dream,' he would say"*: Joan Didion, *The Year of Magical Thinking* (New York: Vintage, 2005), 159.

Dunne died: Eric Homberger, "Obituary: John Gregory Dunne," *Guardian,* January 2, 2004, https://www.theguardian.com/news/2004/jan/02/guardianobituaries.booksobituaries.

Joan didn't dream at all: Didion, *Year of Magical Thinking,* 161.

117 *pattern typical among the bereaved:* Michael Schredl and Hildegard Engelhardt, "Dreaming and Psychopathology: Dream Recall and Dream Content of Psychiatric Inpatients," *Sleep and Hypnosis* 3, no. 1 (2001): 44.

Patricia Garfield's father died: Patricia Garfield, "Dreams in Bereavement," in *Trauma and Dreams,* 186–211.

"All that peered out from the shroud": Philip Roth, *Patrimony: A True Story* (New York: Vintage, 1996), 237.

118 *cared for her grandmother:* Deirdre Barrett, "Through a Glass Darkly: Images of the Dead in Dreams," *Omega: Journal of Death and Dying* 24, no. 2 (1992): 97–108, doi: 10.2190/H9G7-7AK5-15TF-2AWA.

"In the first years following my mother's death": Daphne Merkin, *This Close to Happy: A Reckoning with Depression* (New York: Farrar, Straus and Giroux, 2017), 223.

119 *Socrates entertained:* Alice van Harten, "Socrates on Life and Death (Plato, *Apology* 40C5–41C7)," *Cambridge Classical Journal* 57 (2011): 165–83, doi: 10.1017/S1750270500001317.

According to rabbinic thought: Yisroel Cotlar, "The Kabbalah of Sleep," Chabad-Lubavitch Media Center, http://www.chabad.org/library/article_cdo/aid/559460/jewish/The-Kabbalah-of-Sleep.htm.

"people were always moved to record": Burstein, *Lincoln Dreamt He Died,* 242.

The last dreams of criminals: Author's interview with Shane McCorristine, November 5, 2015.

Visitation dreams: Kelly Bulkeley and Patricia Bulkley, *Dreaming Beyond Death: A Guide to Pre-Death Dreams and Visions* (Boston: Beacon Press, 2006), 18.

hospice patients in a 2014 study: Christopher Kerr et al., "End-of-Life Dreams and Visions: A Longitudinal Study of Hospice Patients' Experiences," *Journal of Palliative Medicine* 17, no. 3 (2014): 296, doi: 10.1089/jpm.2013.0371.

120 *residents of Polish retirement homes:* Wojciech Owczarski, "Therapeutic Effects of the Dreams of Nursing Home Residents in Poland," *Dreaming* 24, no. 4 (2014): 270–78.

hanging out with his own relatives at a shivah: Katie Roiphe, *The Violet Hour* (New York: Random House, 2016), 233.

121 *Sendak had a vivid dream:* Ibid., 251–52.

A man named Bill: Bulkeley and Bulkley, *Dreaming Beyond Death,* 1–3.

women's dreams about the singer Madonna: Kay Turner, *I Dream of Madonna: Women's Dreams of the Goddess of Pop* (San Francisco: Collins Publishers, 1993).

Margie: Ibid., 58.

Chris: Ibid., 83.

122 *Ernest Hartmann, a psychiatry professor:* Ernest Hartmann, "Outline for a Theory on the Nature and Functions of Dreaming," *Dreaming* 6 (1996): 147–54, doi: 10.1037/h0094452.

emerged from the subway on September 11: Deirdre Barrett, "Night Wars," paper presented at the International Association for the Study of Dreams, Tufts University, May 2002.

In PTSD, this system malfunctions: Daniel Freeman and Jason Freeman, "Dispelling the Nightmares of Post-Traumatic Stress Disorder," *Guardian,* April 17, 2014, https://www.theguardian.com/science/blog/2014/apr/17/post-traumatic-stress-disorder-ptsd-cbt.

Alan Siegel: Alan Siegel, "Dreams of Firestorm Survivors," in *Trauma and Dreams,* 159–76.

123 *Kathryn Belicki:* Angela DeDonato, Kathryn Belicki, and Marion Cuddy, "Raters' Ability to Identify Individuals Reporting Sexual Abuse from Nightmare Content," *Dreaming* 6, no. 1 (1996): 33–41, doi: 10.1037/h0094444.

nearly twice as many nightmares: Marion Cuddy and Kathryn Belicki, "Nightmare Frequency and Related Sleep Disturbance as Indicators of a History of Sexual Abuse," *Dreaming* 2, no. 1 (1992): 15, doi: 10.1037/h0094344.

124 *at least an element of recent experience:* Erin Wamsley and Robert Stickgold, "Dreaming and Offline Memory Processing," *Current Biology* 20, no. 23 (2010): R1010, doi: 10.1016/j.cub.2010.10.045.

Children in war zones: Raija-Leena Punamaki et al., "Trauma, Dreaming, and Psychological Distress Among Kurdish Children," *Dreaming* 15, no. 3 (2005): 178.

Sports-studies majors: Daniel Erlacher and Michael Schredl, "Dreams Reflect Waking Sport Activities: A Comparison of Sport and Psychology Students," *International Journal of Sport Psychology* 35, no. 4 (2004): 301.

late 1960s: Edward Tauber et al., "The Effects of Longstanding Perceptual Alterations on the Hallucinatory Content of Dreams," *Psychophysiology* 5, no. 2 (1968): 219.

ethical guidelines: "IRBS: A Brief History," Brandeis University, http://www.brandeis.edu/ora/compliance/irb/101/history.html.

assumed a pinkish tone: Constance Bowe-Anders et al., "Effects of Goggle-Altered Color Perception on Sleep," *Perceptual and Motor Skills* 38, no. 1 (1974): 191–98.

often disturbing: Peter Wortsman, "Howard Roffwarg: A Scientific Champion of Sleep," *Journal of the College of Physicians and Surgeons of Columbia University* 19, no. 2 (1999), http://www.cumc.columbia.edu/psjournal/archive/archives/jour_v19no2/profile.html.

125 *intense nightmares:* Rock, *The Mind at Night*, 92–93.

"saturated the world of Civil War Americans": White, *Midnight in America*, xvi.

"More than anything else": Ibid., xxiii, 34.

Austrian psychiatrist Viktor Frankl: Viktor Frankl, *Man's Search for Meaning: An Introduction to Logotherapy*, trans. Ilse Lasch (New York: Washington Square Press, 1960), 40.

126 *Owczarski rummaged:* Wojciech Owczarski, "Therapeutic Dreams in Auschwitz," *Jednak Ksiazki* 6 (2016): 85–92.

Romanian psychologist Ioana Cosman: Ioana Cosman et al., "Rational and Irrational Beliefs and Coping Strategies Among Transylvanian

Holocaust Survivors: An Exploratory Analysis," *Journal of Loss and Trauma* 18, no. 2 (2013): 179–94, doi: 10.1080/15325024.2012.687322.

127 *Gaetan de Lavilléon:* Gaetan de Lavilléon et al., "Explicit Memory Creation During Sleep Demonstrates a Causal Role of Place Cells in Navigation," *Nature Neuroscience* 18 (2015): 493–95.

Karim Benchenane: Anna Azvolinsky, "Modifying Memories During Sleep," *Scientist,* March 9, 2015, https://www.the-scientist.com/?articles.view/articleNo/42362/title/Modifying-Memories-During-Sleep/.

128 *Garfield advised:* Patricia Garfield, *The Healing Power of Dreams* (New York: Simon and Schuster, 1991), 187.

7. Nightmares

131 *peak around age ten:* Michael Schredl et al., "Longitudinal Study of Nightmares in Children: Stability and Effect of Emotional Symptoms," *Child Psychiatry and Human Development* 40, no. 3 (2009): 439–49.

four out of five adults could recall: Brant Hasler and Anne Germain, "Correlates and Treatments of Nightmares in Adults," *Sleep Medicine Clinics* 4, no. 4 (2009): 507, doi: 10.1016/j.jsmc.2009.07.012.

most common nightmare scenario: Michael Schredl, "Nightmare Frequency and Nightmare Topics in a Representative German Sample," *European Archives of Psychiatry and Clinical Neuroscience* 260, no. 8 (2010): 565, doi: 10.1007/s00406-010-0112-3.

132 *impact of dreams on romantic relationships:* Dylan Selterman et al., "Dreaming of You: Behavior and Emotion in Dreams of Significant Others Predict Subsequent Relational Behavior," *Social Psychological and Personality Science* 5, no. 1 (2013): 111, doi: 10.1177/1948550613486678.

"Dream delusions": Erin Wamsley et al., "Delusional Confusion of Dreaming and Reality in Narcolepsy," *Sleep* 37, no. 2 (2014): 419–22, doi: 10.5665/sleep.3428.

133 *"I remember seeing his arms stretch out":* Author's interview with Julie Flygare, March 30, 2017.

Mehmet Agargun: Mehmet Yucel Agargun et al., "Case Report: Nightmares Associated with the Onset of Mania: Three Case Reports," *Sleep and Hypnosis* 5, no. 4 (2003): 192–96.

134 *In the 1970s, psychologist Joseph De Koninck:* Joseph M. De Koninck and

David Koulack, "Dream Content and Adaptation to a Stressful Situation," *Journal of Abnormal Psychology* 84, no. 3 (1975): 250–60, doi: 10.1037/h0076648.

PTSD sufferer David Morris: David Morris, *The Evil Hours: A Biography of Post-Traumatic Stress Disorder* (Boston: Houghton Mifflin Harcourt, 2015), 119.

135 *Adolescents in rural China:* Xianchen Liu, "Sleep and Adolescent Suicidal Behavior," *Sleep* 27, no. 7 (2014): 1351, doi: 10.1093/sleep/27.7.1351.

more than thirty-six thousand Finnish adults: Antti Tanskanen et al., "Nightmares as Predictors of Suicide," *Sleep* 24, no. 7 (2001): 844.

American Indian Zuni: Barbara Tedlock, "Zuni and Quiché Dream Sharing and Interpreting," in *Dreaming: Anthropological and Psychological Interpretations,* 106.

Rarámuri of northwestern Mexico: Merrill, "The Rarámuri Stereotype of Dreams," 200.

"If your dream contains danger": Author's interview with Jean Kim, February 2, 2017.

psychologist Gail Heather-Greener: Gail Heather-Greener et al., "An Investigation of the Manifest Dream Content Associated with Migraine Headaches: A Study of the Dreams That Precede Nocturnal Migraines," *Psychotherapy and Psychosomatics* 65, no. 4 (1996): 216.

136 *burn victims:* Isabelle Raymond et al., "Incorporation of Pain in Dreams of Hospitalized Burn Victims," *Sleep* 25, no. 7 (2002): 765–70.

cases of seemingly healthy people: Malvinder S. Parmar and Alejandro F. Luque-Coqui, "Killer Dreams," *Canadian Journal of Cardiology* 14, no. 11 (1998): 1389–91.

137 *dying in their sleep:* Shelley R. Adler, *Sleep Paralysis: Night-Mares, Nocebos, and the Mind-Body Connection* (New Brunswick, NJ: Rutgers University Press, 2011).

"We did an autopsy in each case": Wayne King, "Nightmares Suspected in Bed Deaths of 18 Laotians," *New York Times,* May 10, 1981, http://www.nytimes.com/1981/05/10/us/nightmares-suspected-in-bed-deaths-of-18-laotians.html.

"like a thin layer of butter": Adler, *Sleep Paralysis,* 103.

139 *Imagery rehearsal therapy:* Barry Krakow and Antonio Zadra, "Imagery

Rehearsal Therapy: Principles and Practice," *Sleep Medicine Clinics* 5, no. 2 (2010): 289–98, doi: 10.1016/j.jsmc.2010.01.004.

process is so unpleasant: Melynda D. Casement and Leslie M. Swanson, "A Meta-Analysis of Imagery Rehearsal for Post-Trauma Nightmares: Effects on Nightmare Frequency, Sleep Quality, and Posttraumatic Stress," *Clinical Psychology Review* 32, no. 6 (2012): 566–74, doi: 10.1016/j.cpr.2012.06.002.

rehashing traumatic nightmares: M. Lu et al., "Imagery Rehearsal Therapy for Posttraumatic Nightmares in U.S. Veterans," *Journal of Traumatic Stress* 22, no. 3 (2009): 236–39, doi: 10.1002/jts.20407.

140 *Skip Rizzo:* Author's interview with Skip Rizzo, May 25, 2017.

143 *phobic patients said they would rather be treated:* A. Garcia-Palacios et al., "Comparing Acceptance and Refusal Rates of Virtual Reality Exposure vs. In Vivo Exposure by Patients with Specific Phobias," *CyberPsychology and Behavior* 10, no. 5 (2007): 722, doi: 10.1089/cpb.2007.9962.

psychologists at Utrecht University: V. I. Spoormaker and J. van den Bout, "Lucid Dreaming Treatment for Nightmares: A Pilot Study," *Psychotherapy and Psychosomatics* 75, no. 6 (2006): 389–94, doi: 10.1159/000095446.

144 *Steve Volk was plagued:* Matt Kielty, "Wake Up and Dream," *Radiolab,* WNYC, January 23, 2012, http://www.radiolab.org/story/182747-wake-up-dream/.

8. Diagnosis

146 *"They started to do fancy artwork":* Author's interview with Robert Stickgold, March 13, 2017.

147 *lied in session:* Matt Blanchard and Barry A. Farber, "Lying in Psychotherapy: Why and What Clients Don't Tell Their Therapist About Therapy and Their Relationship," *Counselling Psychology Quarterly* 29, no. 1 (2016): 90, doi: 10.1080/09515070.2015.1085365.

Some languages, like Greek: Mark Blechner, *The Dream Frontier* (New York: Routledge, 2001), 42.

148 *"not imputed to him as a sin":* Saint Thomas Aquinas, *Summa Theologica Part II ("Secunda Secundae"),* trans. Fathers of the English Dominican Province (New York: Benziger Brothers, 1971).

traditionally undergo purification: Matt Goldish, *Jewish Questions: Responsa on Sephardic Life in the Early Modern Period* (Princeton, NJ: Princeton University Press, 2008), 134.

the high priest: Elon Gilad, "Ten Things You Probably Don't Know About Yom Kippur," *HaAretz,* September 10, 2013, https://www.haaretz.com/jewish/10-things-you-didn-t-know-about-kippur-1.5332142.

The Examined Life: Stephen Grosz, *The Examined Life: How We Lose and Find Ourselves* (New York: W. W. Norton, 2013), 136–40.

149 *suppress negative feelings:* Josie Malinowski, "Dreaming and Personality: Wake-Dream Continuity, Thought Suppression, and the Big Five Inventory," *Consciousness and Cognition* 38 (2015): 9, doi: 10.1016/j.concog.2015.10.004.

dreams churn up topics: Daniel Wegner et al., "Dream Rebound: The Return of Suppressed Thoughts in Dreams," *Psychological Science* 15, no. 4 (2004): 232, doi: 10.1111/j.0963-7214.2004.00657.x.

white bears: Daniel Wegner, "Paradoxical Effects of Thought Suppression," *Journal of Personality and Social Psychology* 53, no. 1 (1987): 5–13.

150 *avoid thoughts of cigarettes:* James A. Erskine et al., "I Suppress, Therefore I Smoke: Effects of Thought Suppression on Smoking Behavior," *Psychological Science* 21, no. 9 (2010): 1225, doi: 10.1177/0956797610378687.

suppress images of chocolate: Janet Polivy et al., "The Effect of Deprivation on Food Cravings and Eating Behavior in Restrained and Unrestrained Eaters," *Eating Disorders* 38, no. 4 (2005): 301, doi: 10.1002/eat.20195.

try too hard to think positive: Richard M. Wenzlaff and David D. Luxton, "The Role of Thought Suppression in Depressive Rumination," *Cognitive Therapy and Research* 27, no. 3 (2003): 293, doi: 10.1023/A:1023966400540.

neuropsychoanalysis: Casey Schwartz, "When Freud Meets fMRI," *Atlantic,* August 25, 2015, https://www.theatlantic.com/health/archive/2015/08/neuroscience-psychoanalysis-casey-schwartz-mind-fields/401999/.

damage to the pontine brain stem: Mark Solms, "Dreaming and REM Sleep Are Controlled by Different Brain Mechanisms," *Behavioral and Brain Sciences* 23, no. 6 (2000): 843.

151 *emotional-processing systems:* Solms, *The Brain and the Inner World,* 112–32.

enjoyed "a massive increase": Ibid., 207.

152 *"Cultural stereotypes portray":* William Domhoff, "The Repetition of Dreams and Dream Elements: A Possible Clue to a Function of Dreams?," in *The Functions of Dreaming,* ed. Alan Moffitt (Albany: State University of New York Press, 1993), 293–320.

The Individual and His Dreams: Calvin Hall and Vernon Nordby, *The Individual and His Dreams* (New York: New American Library, 1972), 82.

Antonio Zadra: Antonio Zadra, "Recurrent Dreams: Their Relation to Life Events," in *Trauma and Dreams,* 231–47.

experiencing recurrent dreams: P. R. Robbins and F. Houshi, "Some Observations on Recurrent Dreams," *Bulletin of the Menninger Clinic* 47, no. 3 (1983): 262.

better mental health: Ronald Brown and Don Donderi, "Dream Content and Self-Reported Well-Being Among Recurrent Dreamers, Past-Recurrent Dreamers, and Nonrecurrent Dreamers," *Journal of Personality and Social Psychology* 50 (1986): 612–23.

"forced to exercise their mental muscles": Van de Castle, *Our Dreaming Mind,* 342.

153 *Kelsey Osgood:* Kelsey Osgood, *How to Disappear Completely: On Modern Anorexia* (New York: Overlook Press, 2014).

"One was about cereal": Author's interview with Kelsey Osgood, May 8, 2016.

half of bulimic patients: Christoph Lauer and Jurgen-Christian Krieg, "Sleep in Eating Disorders," *Sleep Medicine Reviews* 8, no. 2 (2004): 109.

1 percent of the dreams: Tore Nielsen and Russell Powell, "Dreams of the *Rarebit Fiend:* Food and Diet as Instigators of Bizarre and Disturbing Dreams," *Frontiers in Psychology* 6, no. 47 (2015), doi: 10.3389/fpsyg.2015.00047.

154 *culinary desert:* Tore Nielsen et al., "The Typical Dreams of Canadian University Students," *Dreaming* 13, no. 4 (2003): 211–35, doi: 10.1023/B:DREM.0000003144.40929.0b.

first weeks of abstinence: George Christo and Christine Franey, "Addicts' Drug-Related Dreams: Their Frequency and Relationship to Six-

Month Outcomes," *Substance Use and Misuse* 31, no. 1 (1996): 1–15, doi: 10.3109/10826089609045795.

Blackout: Sarah Hepola, *Blackout: Remembering the Things I Drank to Forget* (New York: Grand Central Publishing, 2015).

"I'd be at a party": Author's interview with Sarah Hepola, May 9, 2016.

155 *warning sign:* Claudio Colace, *Drug Dreams: Clinical and Research Implications of Dreams About Drugs in Drug-Addicted Patients* (London: Routledge, 2014), 57.

"Gregor Samsa": Franz Kafka, *The Metamorphosis,* trans. Edwin Muir and Willa Muir (New York: Random House, 1933).

bipolar disorder: Kathleen Beauchemin and Peter Hays, "Prevailing Mood, Mood Changes and Dreams in Bipolar Disorder," *Journal of Affective Disorders* 35, no. 1 (1995): 41–49, doi: 10.1016/0165-0327(95)00036-M.

156 *blood tests:* Sally Adee, "Suicidal Behaviour Predicted by Blood Test Showing Gene Changes," *New Scientist,* August 19, 2015, https://www.newscientist.com/article/mg22730354-000-suicidal-behaviour-predicted-by-blood-test-showing-gene-changes/.

app-based algorithms: "Predicting Suicide Risk," *News in Health,* October 2015, https://newsinhealth.nih.gov/2015/10/predicting-suicide-risk.

Both groups' dreams dwell on death: Myron Glucksman and Milton Kramer, "Manifest Dream Content as a Predictor of Suicidality," *Psychodynamic Psychiatry* 45, no. 2 (2017): 175–85, doi: 10.1521/pdps.2017.45.2.175.

157 *"suicide is self-murder":* Author's interview with Myron Glucksman, July 14, 2017.

Dreaming: An Opportunity for Change: Myron L. Glucksman, *Dreaming: An Opportunity for Change* (New York: Rowman and Littlefield, 2006), 146.

158 Running on Empty: Carrie Arnold, *Running on Empty: A Diary of Anorexia and Recovery* (Livonia, MI: First Page, 2004).

"literally drooling": Author's interview with Carrie Arnold, May 10, 2016.

William Dement smoked: Stanley Krippner et al., *Extraordinary Dreams and How to Work with Them* (Albany: State University of New York Press, 2001), 70.

159 *most recent fever dream:* Michael Schredl et al., "Bizarreness in Fever

Dreams: A Questionnaire Study," *International Journal of Dream Research* 9, no. 1 (2016): 86.

"Fever can be associated with delirium": Author's interview with Jean Kim, February 2, 2017.

"You're doing enormous sensory processing": Author's interview with Patrick McNamara, February 24, 2017.

160 *Vasily Kasatkin:* Barrett, *The Committee of Sleep,* 134–35.

"with a very unpleasant": Vasily Kasatkin, *A Theory of Dreams,* trans. Susanne van Doorn (N.P.: Lulu, 2014), 22.

"Often, unpleasant dreams appeared": Ibid., 24.

"I soon got into my stride": Oliver Sacks, *A Leg to Stand On* (New York: Touchstone, 1984), 1.

161 *"I had lost the inner image":* Ibid., 54.

dreams of being frozen: Barrett, *The Committee of Sleep,* 128.

162 *dreams about being paralyzed:* J. G. MacFarlane and T. L. Wilson, "A Relationship Between Nightmare Content and Somatic Stimuli in a Sleep-Disordered Population: A Preliminary Study," *Dreaming* 16, no. 1 (2006): 53, doi: 10.1037/1053-0797.16.1.53.

doctors at Britain's Swansea University: Samantha Fisher et al., "Emotional Content of Dreams in Obstructive Sleep Apnea Hypopnea Syndrome Patients and Sleepy Snorers Attending a Sleep-Disordered Breathing Clinic," *Journal of Clinical Sleep Medicine* 7, no. 1 (2011): 69.

REM sleep behavior disorder: Dimitri Markov et al., "Update on Parasomnias: A Review for Psychiatric Practice," *Psychiatry* 3, no. 7 (2006): 69–76.

stories of RBD patients: Megan McCann, "While Asleep, Some People Act Out Their Dreams," Northwestern Memorial Hospital, press release, https://www.nm.org/about-us/northwestern-medicine-newsroom/press-releases/2012/while-asleep-some-people-act-out-their-dreams.

Carlos Schenck: Carlos Schenck et al., "Delayed Emergence of a Parkinsonian Disorder or Dementia in 81% of Older Men Initially Diagnosed with Idiopathic Rapid Eye Movement Sleep Behavior Disorder: A 16-Year Update on a Previously Reported Series," *Sleep Medicine* 14, no. 8 (2013): 744–48, doi: 10.1016/j.sleep.2012.10.009.

163 *50 percent chance:* Ronald Potsuma et al., "Parkinson Risk in Idiopathic REM Sleep Behavior Disorder," *Neurology* 84, no. 11 (2015): 1104–13, doi: 10.1212/WNL.0000000000001364.

Lewy bodies: Daniel Claassen and Scott Kutscher, "Sleep Disturbances in Parkinson's Disease Patients and Management Options," *Nature and Science of Sleep* 3 (2011): 125–33, doi: 10.2147/NSS.S18897.

Robert Bosnak: Robert Bosnak, "Integration and Ambivalence in Transplants," in *Trauma and Dreams,* 217–30.

164 *"There are levels of awareness":* Author's interview with Rebecca Fenwick, January 18, 2017.

9. Dream Groups

168 *"free and spontaneous":* Montague Ullman and Nan Zimmerman, *Working with Dreams: Self-Understanding, Problem-Solving, and Enriched Creativity Through Dream Appreciation* (Los Angeles: Jeremy P. Tarcher, 1979), 217.

169 *"A primary reason":* Taylor, *The Wisdom of Your Dreams,* 62.

170 *Shane McCorristine:* Author's interview with Shane McCorristine, November 5, 2015.

apologizing for the topic: Dan Piepenbring, "The Enthralling, Anxious World of Vladimir Nabokov's Dreams," *New Yorker,* February 8, 2018, https://www.newyorker.com/books/page-turner/what-vladimir-nabokov-saw-in-his-dreams.

171 *Sarah Koenig:* "511: The Seven Things You're Not Supposed to Talk About," *This American Life,* NPR (in collaboration with Chicago Public Media), November 8, 2013, transcript, https://www.thisamericanlife.org/511/the-seven-things-youre-not-supposed-to-talk-about.

Charlie Brooker: Charlie Brooker, "Other People's Dreams Are Boring—Why Would We Want a Machine that Can Record That?," *Guardian,* November 1, 2010, https://www.theguardian.com/commentisfree/2010/nov/01/charlie-brooker-dream-recording-machine-inception.

Michael Chabon: Michael Chabon, "Why I Hate Dreams," *Daily* (blog), *New York Review of Books,* June 15, 2012, http://www.nybooks.com/daily/2012/06/15/why-i-hate-dreams/.

"Tellers of dreams have some basic obstacles": Author's interview with James Phelan, July 28, 2016.

172 *"listener who inherently doesn't really care":* Author's interview with Alison Booth, July 25, 2016.

"it's got to be very lenient": Author's interview with Robert Stickgold, March 13, 2017.

173 *Victims of abuse:* Meg Jay, *Supernormal: The Untold Story of Resilience* (New York: Hachette, 2017), 241.

"the act of not discussing": James W. Pennebaker, quoted in ibid.

dream-group model: Montague Ullman, *Appreciating Dreams: A Group Approach* (Thousand Oaks, CA: Sage Publications, 1996).

"I experienced a sense of diminishing returns": Ullman and Zimmerman, *Working with Dreams*, 10.

"Trust, communion, and a sense of solidarity": Ibid., 257.

174 *a website devoted to Ullman's legacy:* William R. Stimson, "The Worldwide Montague Ullman Dream Group," http://www.billstimson.com/dream_group/dream_links.htm.

Mark Blagrove learned: Author's interview with Mark Blagrove, June 2016.

175 *personal insight after they shared:* Christopher Edwards et al., "Comparing Personal Insight Gains Due to Consideration of a Recent Dream and Consideration of a Recent Event Using the Ullman and Schredl Dream Group Methods," *Frontiers in Psychology* 6 (2015): 831, doi: 10.3389/fpsyg.2015.00831.

help people improve a relationship: Clara E. Hill and Dana R. Falk, "The Effectiveness of Dream Interpretation Groups for Women Undergoing a Divorce Transition," *Dreaming* 5, no. 1 (1995): 29.

dreams could help couples communicate: Misty R. Kolchakian and Clara E. Hill, "Dream Interpretation with Heterosexual Dating Couples," *Dreaming* 12, no. 1 (2002): 1–16, doi: 10.1023/A:1013884804836.

176 *"Every morning we would start":* Wojciech Owczarski, "The Ritual of Dream Interpretation in the Auschwitz Concentration Camp," *Dreaming* 27, no. 4 (2017): 278, doi: 10.1037/drm0000064.

177 *social worker Susan Hendricks:* Author's interview with Susan Hendricks, January 17, 2017.

10. Control

187 *"Before having yourself tasted such delight"*: van Eeden, *The Bride of Dreams,* 170.

190 *"box-based architecture"*: "Self-Care Lotus Retreat," Kalani Honua, https://kalani.com/retreat/self-care-lotus-retreat-2/.

"I was just so in awe": Author's interview with Kristen LaMarca, October 2016.

191 *Nolan said:* Ashley Lee, "Christopher Nolan Talks *Inception* Ending, Batman, and 'Chasing Reality' in Princeton Grad Speech," *Hollywood Reporter,* June 1, 2015, https://www.hollywoodreporter.com/news/christopher-nolan-princeton-graduation-speech-799121.

Reddit: https://www.reddit.com/r/LucidDreaming/.

194 *Effective reality tests:* LaBerge and Rheingold, *Exploring the World of Lucid Dreaming,* 59–65.

196 *dream signs:* Ibid., 40–47.

"some sort of super-protozoan": Ibid., 42.

197 *regular meditators had clearer memories:* Henry Reed, "Improved Dream Recall Associated with Meditation," *Journal of Clinical Psychology* 34, no. 1 (1978): 150, doi: 10.1002/1097-4679(197801)34:1<150::AID-JCLP2270340133>3.0.CO;2-1.

Jayne Gackenbach also noticed: Jayne Gackenbach et al., "Lucid Dreaming, Witnessing Dreaming, and the Transcendental Meditation Technique: A Developmental Relationship," *Lucidity Letter* 5, no. 2 (1986): 3.

198 *"If you keep the mind sufficiently active":* LaBerge and Rheingold, *Exploring the World of Lucid Dreaming,* 95.

"Try to observe the images": Ibid., 98–99.

"Just being engaged": LaBerge, presentation at Dreaming and Awakening.

199 *appears to be most common in children and adolescents:* David Saunders et al., "Lucid Dreaming Incidence: A Quality Effects Meta-Analysis of 50 Years of Research," *Consciousness and Cognition* 43 (2016): 197–215, doi: 10.1016/j.concog.2016.06.002.

six- and seven-year-olds: Ursula Voss et al., "Lucid Dreaming: An Age-Dependent Brain Dissociation," *Journal of Sleep Research* 21, no. 6 (2012): 634–42, doi: 10.1111/j.1365-2869.2012.01022.x.

"need for cognition": Mark Blagrove and S. J. Hartnell, "Lucid Dreaming: Associations with Internal Locus of Control, Need for Cognition and Creativity," *Personality and Individual Differences* 28 (2000): 41–47.

lucidity and creativity: Tadas Stumbrys and Michael Daniels, "An Exploratory Study of Creative Problem Solving in Lucid Dreams: Preliminary Findings and Methodological Considerations," *International Journal of Dream Research* 3, no. 2 (2010): 121–29; Daniel Bernstein and Kathryn Belicki, "On the Psychometric Properties of Retrospective Dream Content Questionnaires," *Imagination, Cognition and Personality* 15, no. 4 (1996): 351–64.

gamers and lucid dreamers: Jayne Gackenbach and Harry T. Hunt, "A Deeper Inquiry into Lucid Dreams and Video Game Play," in *Lucid Dreaming: New Perspectives on Consciousness in Sleep,* eds. Ryan Hurd and Kelly Bulkeley (Santa Barbara, CA: Praeger, 2014), 235.

"The major parallel between gaming and dreaming": Jayne Gackenbach, quoted in Katie Drummond, "Video Games Change the Way You Dream," *Verge,* January 21, 2014, https://www.theverge.com/2014/1/21/5330636/video-games-effect-on-dreams.

200 *involving hundreds of professional athletes:* Daniel Erlacher, Tadas Stumbrys, and Michael Schredl, "Frequency of Lucid Dreams and Lucid Dream Practice in German Athletes," *Imagination, Cognition and Personality* 31, no. 3 (2012): 237, doi: 10.2190/IC.31.3.f.

2016 meta-analysis: Saunders et al., "Lucid Dreaming Incidence," 197.

201 *sifted through the academic literature:* Tadas Stumbrys et al., "Induction of Lucid Dreams: A Systematic Review of Evidence," *Consciousness and Cognition* 3 (2012): 1465–75, doi: 10.1016/j.concog.2012.07.003.

202 *longer, more naturalistic study:* David Saunders et al., "Exploring the Role of Need for Cognition, Field Independence and Locus of Control on the Incidence of Lucid Dreams During a 12-Week Induction Study," *Dreaming* 27, no. 1 (2017): 68, doi: 10.1037/drm0000044.

Australian study: D. J. Aspy et al., "Reality Testing and the Mnemonic Induction of Lucid Dreams: Findings from the National Australian Lucid Dream Induction Study," *Dreaming* 27, no. 3 (2017): 206–31, doi: 10.1037/drm0000059.

203 *galantamine reduces "REM sleep latency":* Meir H. Kryger et al., *Principles*

and Practice of Sleep Medicine, 5th ed. (St. Louis: Elsevier Saunders, 2010), 1530.

204 *nineteen lucid dreamers who incorporated galantamine:* Gregory Scott Sparrow et al., "Assessing the Perceived Differences in Post-Galantamine Lucid Dreams vs. non-Galantamine Lucid Dreams," *International Journal of Dream Research* 9, no. 1 (2016): 71.

205 *Ursula Voss:* Ursula Voss et al., "Lucid Dreaming: A State of Consciousness with Features of Both Waking and Non-Lucid Dreaming," *Sleep* 32, no. 9 (2009): 1191.

scanning hypothesis: Dement, *Some Must Watch,* 47–52.

People who lose their sight: Craig S. Hurovitz et al., "The Dreams of Blind Men and Women: A Replication and Extension of Previous Findings," *Dreaming* 9 (1999): 183.

206 *Peretz Lavie:* Peretz Lavie, *The Enchanted World of Sleep* (New Haven, CT: Yale University Press, 1996), 87.

"Longer durations in lucid dreams": Daniel Erlacher et al., "Time for Actions in Lucid Dreams: Effects of Task Modality, Length, and Complexity," *Frontiers in Psychology* 4 (2013): 1013, doi: 10.3389/fpsyg.2013.01013.

squeeze their fists: Martin Dresler et al., "Neural Correlates of Dream Lucidity Obtained from Contrasting Lucid versus Non-Lucid REM Sleep: A Combined EEG/fMRI Case Study," *Sleep* 35, no. 7 (2012): 1017, doi: 10.5665/sleep.1974.

207 *"dreams are an excellent means":* Author's interview with Katja Valli, June 2016.

resemblance to the psychotic brain: Martin Dresler et al., "Neural Correlates of Insight in Dreaming and Psychosis," *Sleep Medicine Reviews* 20 (2015): 92–99, doi: 10.1016/j.smrv.2014.06.004.

Lucid dreaming can also help people: Adhip Rawal, "Could We One Day Heal the Mind by Taking Control of Our Dreams?," *Conversation,* July 14, 2016, https://theconversation.com/could-we-one-day-heal-the-mind-by-taking-control-of-our-dreams-60886.

German psychologist Paul Tholey: LaBerge, *Lucid Dreaming.*

208 *"had a liberating and encouraging effect":* Paul Tholey, "A Model for Lucidity Training as a Means of Self-Healing and Psychological Growth,"

in *Conscious Mind, Sleeping Brain: Perspectives on Lucid Dreaming,* eds. Jayne Gackenbach and Stephen LaBerge (New York: Springer, 1988), 265.

209 *"They were in the driver's seat":* Author's interview with Line Salvesen, July 2, 2017.

210 *a tool in performance and exercise:* Daniel Erlacher, "Practicing in Dreams Can Improve Your Performance," *Harvard Business Review,* April 2012, https://hbr.org/2012/04/practicing-in-dreams-can-improve-your-performance.

Epilogue: My Night Life

213 *Japanese researchers:* T. Horikawa et al., "Neural Decoding of Visual Imagery During Sleep," *Science* 340 (2013): 639, doi: 10.1126/science.1234330.

"answers will come from rodent models": Author's interview with Matt Wilson, March 16, 2017.

Gaetan de Lavilléon: Lavilléon, "Explicit Memory Creation During Sleep."

215 *Rubin Naiman:* Naiman, "Dreamless," 77.

INDEX